3 2109 00460 4568

9.95
55P

P9-CLC-057

HOW THE
MILITARY
WILL HELP
YOU PAY FOR
COLLEGE

WITHDRAWAL

HOW THE MILITARY WILL HELP YOU PAY FOR COLLEGE

THE HIGH SCHOOL STUDENT'S GUIDE TO ROTC, THE ACADEMIES, AND SPECIAL PROGRAMS

SECOND EDITION

DON M. BETTERTON

Peterson's Guides

Princeton, New Jersey

Copyright © 1990 by Peterson's Guides, Inc.

Previous edition copyright 1985

All rights reserved. No part of this book may be reproduced, stored in a retrieval system, or transmitted, in any form or by any means—electronic, mechanical, photocopying, recording, or otherwise—except for citations of data for scholarly or reference purposes with full acknowledgment of title, edition, and publisher and written notification to Peterson's Guides prior to such use.

Library of Congress Cataloging-in-Publication Data
Betterton, Don M., 1938–
    How the military will help you pay for college : the high school student's guide to ROTC, the academies, and special programs / Don M. Betterton. — 2nd ed.
      p.   cm.
    Summary: Description of the military's financial aid programs for college students, including ROTC, attendance at the service academies, and various special programs.
    ISBN 0-87866-996-5
    1. Soldiers—Education, Non-Military—United States. 2. Scholarships—United States. 3. Student aid—United States. 4. Military education—United States. [1. Scholarships. 2. Student aid. 3. Military education.] I. Title.
U716.B48   1990
378.3'4'0973—dc20   89-29864
                                     CIP

Composition and design by Peterson's Guides

Printed in the United States of America

10 9 8 7 6 5 4 3 2 1

Reference
U
716
B48
1990

Contents

(continued)

Acknowledgments

Among those who helped with the writing of this book, I first would like to thank my wife, Finn, for her encouragement, advice, and countless hours on the word processor. My son Thom also contributed by providing suggestions for design and layout.

In gathering information about the ROTC programs, Colonel Bob Selkis of the Princeton University Army ROTC Unit, Captain Ron Ball of the University of Pennsylvania Naval ROTC Unit, and Captain Ryall Smith of the Rutgers Air Force ROTC Unit were patient and knowledgeable answerers of my many questions. Anita Lancaster from the Office of the Assistant Secretary of Defense helped coordinate the review of the manuscript by the military services. Among those in Washington, D.C., the Army's Major Jim Artis was particularly helpful.

For their assistance with the second edition, I would like to thank the commanding officers of the ROTC units at the University of Pennsylvania, Princeton, and Rutgers; the public information officers at the service academies; and the Army, Navy, Air Force, and Marine Corps recruiting offices in Trenton, New Jersey.

In retrospect, the seemingly straightforward job of collecting descriptions of military college aid programs was a much more ambitious project than I had imagined. There are hundreds of different programs—some created by congressional law, some designed by the Department of Defense, and some run by the individual services. Moreover, the programs are being updated and revised constantly as the military tries to develop the most effective ways to help finance the education of its officers and enlisted servicemembers.

In spite of these complexities, I tried to keep the basic purpose in mind—to put it all in one place, to get it right, and to explain it clearly. Where I've been successful, I readily acknowledge those who have been so helpful. On the other hand, I take the responsibility for any errors of fact or interpretation that may have slipped into this manuscript.

Introduction

Today the cost of going to college is quite high, but the value of a college education is even higher. As the U.S. economy becomes increasingly dependent on technology—in manufacturing, engineering, finance, and communications—your level of skill and your experience become more important. Not only is a college education crucial in enabling you to keep pace with the requirements of a job, it also will mean a higher standard of living for you. Based on a study by the Bureau of the Census, average lifetime earnings are expected to be $985,000 for someone with a high school diploma, compared to $1,400,000 for a college graduate.

While the value of a college education may be somewhat hard to pin down, the cost is all too obvious. For a full-time student entering in the fall of 1989, the average bill for four years is expected to be about $26,000 at a public college and $53,000 at a private institution. Even if you go to college only part-time you will still end up paying a similar amount, although it may be easier on your pocketbook to spread the payments over a longer period of time.

How to come up with the money to meet such an expense is one of the major dilemmas facing nearly every individual or family today. There is an extensive nationwide system of financial aid—totaling $26-billion each year—to help students pay for the portion of college costs they cannot afford. However, while this "need-based" assistance is quite important, it is directed primarily at students from low-income families who attend college full-time. The focus of military financial aid is quite different. The military programs do not consider need but are based on either a payment for training or a reward for service. This large source of money (about $1-billion each year) can prove quite helpful in assisting a wide range of students, from a typical college undergraduate from a middle-class family to an enlisted servicemember who is independent of family support. The military financial aid programs are by far the largest source of college money that is not based on need.

In *How the Military Will Help You Pay for College*, you will learn about the two major ways in which the military provides financial aid. First, as described in Part I of this book, is college money for officer candidates—tuition assistance and monthly pay in return for your promise to serve as an officer in the Army, Navy, Air Force, Marine Corps, Coast Guard, or Merchant Marine. Most of this money is awarded to high school seniors who go directly to college. The main benefits are reduced or free tuition and $100 per month if you are part of the Reserve Officers' Training Corps (ROTC) Scholarship Program or free tuition, room and board, and $500 per month if you enroll at one of the service academies. ROTC units are located on college campuses and provide military training for a few hours each week. The five service academies—West Point, Annapolis, the Air Force Academy, the Coast Guard Academy, and the Merchant Marine Academy—are military establishments where education and training for the Armed Forces are combined. (The Merchant Marine Academy is more accurately a military*like* civilian academy. It has a close connection to the Navy in that it trains officers to serve aboard privately owned merchant ships and may also commission Navy officers. The financial benefits are similar to those offered by the military service academies.) If you are already in college, there is also financial aid for you through ROTC scholarships for enrolled students or special commissioning programs.

The second major category of military aid programs, as described in Part II, is for enlisted servicemembers. This takes the form of tuition aid and monthly payments as a "fringe benefit" in return for serving in the military. If you are thinking of postponing your college plans and joining the military first, knowledge of these sources of aid can be an important part of your decision-making process. While serving on active duty you can receive credit both for learning a military skill and for going to college during off-duty hours. After you have completed your obligated service, you can use the money you have saved under the New G.I. Bill to attend college as a civilian. The military also provides assistance to enlisted servicemembers who qualify for officer training programs.

Before the publication of *How the Military Will Help You Pay for College*, there was no single place where you could go to find out about college money the military pays to officer candidates and enlisted servicemembers. Until now, if you wanted to learn about the requirements and benefits of military college aid, you would have had to make separate inquiries at an Army, Navy, Air Force, Marine Corps, or Coast Guard recruiting station, write to the admissions office at each of the service academies, or visit an Army, Navy, or Air Force ROTC unit at a college. This book brings together valuable information about how you might qualify for your share of the $1-billion in college money that the military spends each year.

Here are the military financial aid programs that are covered in *How the Military Will Help You Pay for College*.

Officer Training

ROTC Tuition Scholarships
Army, Navy/Marine Corps, Air Force

Free Service Academies
West Point, Annapolis, Air Force Academy, Coast Guard Academy,
Merchant Marine Academy

Special Programs
Navy Nuclear Propulsion Officer Candidate Program

Navy Civil Engineering Corps Collegiate Program
Navy Baccalaureate Degree Completion Program
Marine Corps Platoon Leaders Class
Armed Forces Health Professions Scholarship Program
Uniformed Services University of the Health Sciences
Air Force College Senior Engineering Program

Enlisted Educational Benefits

In Service
Free college credit for attending military schools and learning job skills
Tuition reduction for courses taken while off duty

After Service
New G.I. Bill
College aid for officer candidates

How This Book Is Organized

How the Military Will Help You Pay for College is divided into three parts. Part I describes programs and benefits for those who choose the direct officer training paths, while Part II covers college financial aid programs for enlisted servicemembers to use while in the service or later. Part III consists of a number of useful appendixes.

Part I: Going to College First: Scholarship Programs for High School Seniors

Chapter 1, Going to College and Becoming an Officer—Which Route Is Best?, encourages you to look at what the military is like and ask yourself whether you are the type of person who will fit in with the military life-style.

Chapter 2, Reserve Officers' Training Corps (ROTC), explains the selection process, what is expected of you in college, and the types of jobs available after you are commissioned.

In Chapter 3, The Service Academies, you learn about this special kind of education—the admissions process, what life is like at an academy, and career opportunities.

Chapter 4, Special Programs, covers routes to a commission in addition to ROTC and the academies.

Chapter 5, Weighing the Options, offers guidelines about which officer program might be best for you.

Part II: Going into the Military First: College Money for Enlisted Servicemembers

Chapter 6, Enlisting in the Military—Is It for You?, tells you what it is like to be an enlisted member of the Armed Forces.

In Chapter 7, Earning College Credits on Active Duty, you can read about how to advance your college education while you serve in the military.

Chapter 8, Banking College Money While Enlisted: The New G.I. Bill, explains who is eligible for this aid, how you save money, and how the benefits are paid.

Chapter 9, Going from Enlisted Servicemember to Officer, describes commissioning programs for enlisted servicemembers.

Part III: Appendixes

Appendix A is a comprehensive list of all colleges that host any kind of ROTC unit.

Appendixes B, C, and D show, by service, where ROTC units are located.

Appendix E is a schedule of military pay and benefits.

Appendix F is a height and weight chart for officer candidates.

Points to Keep in Mind

The material in this book is current as of September 1989. It includes information about college programs through the 1989–90 academic year. It describes the New G.I. Bill, which became effective on July 1, 1985. However, owing to the nature of these programs, the information given is subject to change.

This book is not intended to be the final authority on rules and regulations or to give you complete details on every option. After you have a better feel for which program meets your needs, you should ask for the latest official publication or visit a recruiting station.

When the time comes for you to attend college, you may find you are short of money even though you have tuition benefits from the military (unless you go to one of the academies). Should this occur, remember there are other types of financial aid available. To learn more about sources of aid beyond the military, check with your high school guidance counselor or a college financial aid officer.

PART I

Going to College First: Scholarship Programs for High School Seniors

This part of *How the Military Will Help You Pay for College* is written primarily for high school students who plan to go to college before serving in the military and who want to know more about the two main sources of military scholarships: tuition assistance at the hundreds of colleges that sponsor ROTC or a free education at one of the five service academies—West Point, Annapolis (where you can also prepare for the Marine Corps), the Air Force Academy, the Coast Guard Academy, and the Merchant Marine Academy. The ROTC scholarships are explained in greater detail, not only because they are considerably more numerous than openings at the academies but also because there are many options for you to be aware of.

Part I will also be of value to you if you are now in college and are unaware of the different types of scholarships the military has for enrolled undergraduates—money that can be crucial in helping you finish the last two or three years of an academic program you have already started. Chapter 2 discusses the two- or three-year "in-college" ROTC scholarships that do not receive the same degree of publicity as the four-year ROTC scholarships or attendance at the service academies. As a result, these two- and three-year awards are often overlooked by college students who need financial aid and are interested in becoming military officers. There is also a chapter on special programs that tells you about other sources of college aid, such as the Navy's Nuclear Propulsion Officer Candidate Program and the Marine Corps Platoon Leaders Class.

The following chapters on officer training programs are intended to give you the information you need to compare your abilities and preferences with the requirements of the military service that is offering the scholarship. This should benefit you by clearing up any misconceptions you may have about what options are available.

The military services will also benefit if students have a better understanding of these programs before they apply. The result—a better match between student and service—should produce more satisfied officers and reduce the attrition that comes from a misunderstanding of program rules and regulations.

There is a further reason to give both high school and college students an easy-to-understand "road map" to officer training programs. The idea of serving as a military officer is an attractive opportunity for thousands of college students. Many potential college graduates look upon a military tour of duty as a good way to learn management skills, exercise leadership, and serve their country at the same time. Even if you decide not to pursue a military career and leave the service after four or five years, the experience you will have gained can prove to be very valuable, in terms of both your personal growth and your attractiveness to employers who look favorably upon job applicants with prior military service.

Chapter 1

Going to College and Becoming an Officer—Which Route Is Best?

Participating in ROTC, attending a service academy, or enrolling in a special program for military commissioning not only allows you to become an officer but also provides a source of financial aid that can enable you to attend a college that you might otherwise be unable to afford. The military teaches you to become an officer and pays you to learn at the same time. Your "salary" can be viewed as the extent to which your education is subsidized.

The Services and Their Programs

There are five military services—the Army, Navy, Air Force, Marine Corps, and Coast Guard—that have the responsibility for defending the United States in time of war. Each of these carries out its basic mission through an organization in which officers direct the work of enlisted personnel. Although it is possible to become a military officer by attending an Officer Candidate School after graduation from college, the vast majority of officers are trained while they are in college, by participating in ROTC, attending a service academy, or taking advantage of a special program like the Marine Corps Platoon Leaders Class. In four of the services—the Army, Navy, Air Force, and Marine Corps—both ROTC and academy options are available. (A future Marine Corps officer joins Naval ROTC or attends Annapolis.) The Coast Guard does not sponsor an ROTC program, instead concentrating its college military training at the Coast Guard Academy. The nonmilitary Merchant Marine, although closely allied with the Navy, provides its own officers from the Merchant Marine Academy; it has no ROTC option. Officer candidates in the Army, Air Force, and Coast Guard are called cadets, while potential Navy, Marine Corps, and Merchant Marine officers are referred to as midshipmen.

Is an Officer Training Program Right for Me?

The fundamental purpose of military scholarship programs is to give money to college students as they go through officer training in return for a commitment to serve in the Armed Forces. Through this method of attracting outstanding young men and women, the military hopes to produce an "entry-level" officer who is well educated—in terms of both academics and the workings of the military itself. You should not try for one of these programs if you have moral or religious reservations about serving your country as a military officer. Also, although the financial benefits may be very important, you should not apply if money is the only reason the program appeals to you. The attraction of the financial benefits should be balanced by feelings that you will seriously consider becoming an officer, you will undertake military training with a positive attitude, and you will be open-minded about your future plans. The typical applicant is a young man or woman who is willing to serve at least four or five years as an officer in return for four years of a good education at little or no cost.

The First Test: Am I the Military Type?

The question of whether or not you are cut out to be in the military is an important one you need to answer at the outset. Rather than trying to picture yourself as a second lieutenant or ensign at age 22 or a senior military officer at age 42, look at yourself now. Take a personal inventory and think about what you are like, how you relate to others, and what kind of organization you want to be part of.

Are you energetic, intelligent, well-rounded, and somewhat athletic? Do you organize your time well? Are you a serious student who gets good grades in precollege courses and has an aptitude for math and science?

Are you outgoing and do you work well with others, both one-on-one and in larger groups? Do you willingly take direction from others? Do you enjoy leadership positions?

Can you deal with a structured and disciplined environment? Do you have strong feelings of patriotism? Are you willing to defend your country in wartime?

If you have answered yes to most of these questions, you possess the kinds of characteristics the military services are interested in, and, even more important, you may be the type of person who can adapt to the military's particular life-style. The military is similar to other large businesses in seeking to hire employees who will fit well with its mission and style. While there are many different types of military officers—ranging from the quiet intellectual to the outgoing athlete—there is, nevertheless, a generally accepted set of characteristics for the average officer. This includes a mixture of certain personal traits with a willingness to be part of a large and very structured organization.

The Second Test: What Kind of Military Training Might Be Best for Me?

Assuming that your personal inventory showed you are at least somewhat the "military type," the next step is to see how you match up with the programs offered by the different services. As a starting point, ask yourself which of the following most closely describes your present outlook.

1. I have firsthand knowledge of the military service. I can picture myself as an officer, perhaps even a career officer. I have experience with discipline, in both taking and giving orders. While in college I plan to major in science or engineering. A top-quality education at very low cost is quite important to me.

2. Military service is of interest to me. I have little direct experience, but I am willing to learn more. I'm not sure whether I'm ready to immerse myself completely in a military environment as a college student. I've done well in math and science, but I may want to major in another field. I can look forward to the prospect of four years of service as an officer before deciding whether to stay on. A tuition scholarship is appealing and would widen my range of college possibilities.

3. I don't have a negative attitude toward the military, but it's not something I know very much about. I might like to give it a look. I'm not sure about studying math and science, as I think my interests may be in other areas. I'm concerned about paying for college, but my parents could see me through at least one or two years.

4. I don't think I'm the military type, but I really haven't thought much about it. I doubt I would go for the discipline. I certainly would not want to commit myself to anything until I've been in college for a few years and can see which way I want to head. I possibly could see myself serving in the military if I could get duty that would match my academic interests. I could use a scholarship, but I plan to seek help through other kinds of financial aid programs.

Which paragraph or paragraphs most nearly describe your attitude toward the military? There is a general correlation between the numbers above and the suggested courses of action below.

1. Think seriously about competing for an appointment to a service academy.

2. Plan to enter the national ROTC four-year scholarship competition.

3. Join an ROTC unit in college and see what the military is like. Scholarship opportunities are available if you decide to stay on.

4. Don't get involved with a military program yet, but keep the service in mind for possible entrance after two years of college—either the two-year ROTC option or one of the special programs described in Chapter 4.

In trying to decide whether the military is right for you, you would be wise to avoid extreme discrepancies between the two lists. If, for example, you feel description number 4 is accurate, it is highly unlikely that a military academy or even the four-year ROTC scholarship is right for you. It would be far wiser to choose the third or fourth course of action, since later on, after you are enrolled in college, you could easily find that some aspects of the military fit with your academic interests and that the military life-style is something you can adapt to. On the other hand, if you feel description number 1 suits you, it is worth your while to pursue either the service academy or ROTC scholarship option when you graduate from high school. If you are this far along in your thinking about a possible future in the military, you might as well try for an officer training program as part of your college experience and take advantage of both the financial benefits the services offer and the head start you will get toward a possible military career.

High School Preparation

To improve your chances of winning a four-year ROTC scholarship or receiving an appointment to a service academy, you should be enrolled in a precollege program in your high school. With the exception of the Coast Guard Academy and the Merchant Marine Academy (their requirements are given in Chapter 3), the services don't require you to take specific subjects. However, the Army, Navy, Air Force, and Marine Corps all state that a good high school curriculum is quite important and suggest the following:

English	4 years
Math (through calculus)	4 years
Foreign language	2 years
Laboratory science	2 years
American history	1 year

It is also important for you to be an active member of your school community, and, if possible, to hold leadership positions in extracurricular activities and sports. If your high school has a Junior ROTC unit, you should join the detachment to improve your chances of being selected for an ROTC scholarship or admitted to a service academy. The services also look with favor upon students who are involved in community and church activities and who find the time to hold down a part-time job.

SAT and ACT

You should be prepared to take the SAT or ACT for entrance into the academies and most other colleges. These tests are used by college admissions offices as one of the measures of the academic potential of a prospective college student. The SAT (Scholastic Aptitude Test) is administered by the College Board and is given about six times a year at locations throughout the country. It consists of a math section and a verbal section, with scores ranging from 200 to 800 on each. The ACT, administered by the American College Testing Program, covers four areas—math, English, reading, and science reasoning—with scores ranging from 1 to 36 on each test. Whether you take the SAT or ACT depends on which is required by the colleges you are applying to; the military scholarship programs accept results from either one.

Profile of a Successful Candidate

John or Jane Doe, the typical winner of a four-year ROTC scholarship or a service academy appointment, exhibits certain kinds of characteristics. (We will refer to John for the sake of convenience.)

John is following a curriculum that includes 4 years of English, 4 years of math, 3 or 4 years of a foreign language, 2 years of laboratory science, and 2 years of history. Some of the courses are at the honors level. He is maintaining a B+ average and ranks in the top 15 percent of his class. On the SAT, he received scores of 610 verbal and 640 math. (If he had taken the ACT, his composite score would have been 28.)

John is a member of the National Honor Society. He holds an office in student government and is a candidate for Boys State. John has a leadership position on the student newspaper and belongs to the debate panel and math club. He participates in varsity athletics and is cocaptain of the basketball team.

John's teachers say he is one of the top all-around students in his class, and he makes a positive contribution to the school environment. He is described as intelligent, industrious, well organized, self-confident, concerned, and emotionally mature.

John is also active outside school. He works 4 hours each week in a local drugstore and volunteers some of his time on weekends to help in the town's program to aid the handicapped.

The service believes that a person like John Doe will do well in college—in both academic and military training—and also will have a high probability of becoming a productive officer after graduation.

Common Characteristics of the Officer Training Programs

Officer Pay and Benefits

As a military officer, you will be paid the standard rate for all members of the Armed Forces of your rank and length of service. In addition to your salary, there are significant fringe benefits, such as free medical care and a generous retirement plan. (See Appendix E for details.)

The Difference Between a Regular and a Reserve Officer

In the descriptions of the various officer training programs, there are references to Regular and Reserve officers. You should be aware of the differences between the two designations.

When you are commissioned as a Regular officer, you are on a career path in the military. In the event that you choose not to serve at least twenty years, you must write a letter asking if you can resign your commission. Such requests are normally accepted once you have completed your minimum service obligation. If you plan to make the military your career, it is a definite advantage to be a Regular officer. Among ROTC scholarship recipients, virtually all Navy midshipmen, about 25 percent of Army cadets, and a small number of Air Force cadets are given Regular commissions.

As a Reserve officer, you contract for a specific term, for example, four years of active duty in the case of an ROTC scholarship. Nearly all Air Force ROTC second lieutenants are in this category, along with about 75 percent of Army ROTC graduates. If you want to remain on active duty after your initial obligation, you must request to sign on for a second term. It is possible to serve for twenty years as a Reserve officer, although nearly all career officers are Regulars, either by initial designation or by "integration" (switching from Reserve to Regular).

There is another category of Reserve officer—those who are assigned to the Reserve Forces rather than to active duty. About 50 percent of the officers who are commissioned through the Army ROTC are given orders to the National Guard or Army Reserve. After attending the Basic Course for six months, these officers join the Reserve Forces to finish their obligated service as "weekend warriors." In this case, the time commitment is seven and a half years in the Reserves, the first five and a half years involving drills one weekend per month and two weeks of active duty per year. During the last two years of obligated service, these officers are transferred to inactive Reserve status, in which drills are not required.

Women Officers

Each of the officer training programs mentioned in Chapters 2, 3, and 4 is open to women. With the exception of differences in height and weight standards (see Appendix F) and lower minimums on the physical fitness test, the eligibility rules, benefits, and obligations are the same for women as they are for men.

There is, however, a difference between men and women when it comes to duty assignments. By Act of Congress, women are prohibited from serving in combat units. To what extent this law restricts the types of jobs you can choose varies from service to service. The Navy and Marine Corps have the highest percentage of positions classified as combat related, the Army and Air Force have somewhat fewer, and the Coast Guard is least restrictive, opening virtually all its jobs to men and women alike.

Regardless of the limits imposed by the combat restriction for women, there has been considerable improvement in the position of women within the military in the last ten years. Depending on the service, women make up between 8 and 20 percent of the officers, and they are gradually but steadily moving into the higher-ranking positions.

Medical Requirements

Each candidate for an ROTC scholarship or an academy appointment must pass a medical examination. These exams are coordinated by the Department of Defense Medical Examination Review Board. Even if you apply for more than one type of scholarship, you will have to take only one physical. Each service arranges the specific medical standards it wants for its own programs with the Medical Examination Review Board.

The general rule about medical standards is that they vary considerably and can be quite complicated. In spite of this, it is worth having an idea of the general medical requirements at the outset, particularly the eyesight and height and weight rules. If you read this information now, you will avoid applying to a program for which you fall short of a basic standard. Each of the program descriptions in Chapters 2 and 3 includes a section on medical requirements. The descriptions are intended not to be all-inclusive but to give you helpful guidance. Also, keep in mind that medical standards change periodically, and some of them may be waived under certain conditions.

Chapter 2

Reserve Officers' Training Corps (ROTC)

The predominant way for a college student to become a military officer is through the Reserve Officers' Training Corps program: about 12,700 officers come out of ROTC programs each year, while the academies produce 2,950 officers. ROTC is offered by the Army, Navy, and Air Force, while students taking the Marine Corps option participate in Naval ROTC. (The Coast Guard and Merchant Marine do not sponsor ROTC programs.)

Each service that has an ROTC program signs an agreement with a number of colleges to host a unit on their campuses. The Army has 315 detachments, the Navy 69, and the Air Force 150. Each of these units has a commanding officer supervising a staff of active-duty officers and enlisted servicemembers who conduct the military training of cadets and midshipmen. This instruction includes regular class periods in which military science is taught, as well as longer drill sessions in which students concentrate on developing leadership qualities through participation in military formations, physical fitness routines, and field exercises.

It is not necessary for you to attend a college that hosts a unit to participate in ROTC. In addition to the 534 ROTC detachments, there are another 1,661 cross-enrollment opportunities. The Army has 1,082, the Navy 124, and the Air Force 455. You may attend any of the colleges that have a cross-enrollment contract and participate in ROTC at the host institution, provided you are accepted into the unit and you are able to arrange your schedule so that you have time to commute to the ROTC classes and drill sessions. (The Army also has 97 extension centers—small branches of host colleges that provide an additional way to participate in Army ROTC.)

As a member of an ROTC unit, you are a part-time cadet or midshipman. You are required to wear a uniform and adhere to military discipline when you attend an ROTC class or drill, but not at other times. Since this involvement averages only about 4 hours per week, most of the time you will enjoy the same life-style as a typical college student. You must realize, however, that while you are an undergraduate you are being trained to become an officer when you graduate. You therefore will have a number of obligations and responsibilities your classmates do not face. Nevertheless, the part-time nature of your military training is the major difference between participating in ROTC and enrolling at a service academy, where you are in a military environment 24 hours a day.

In each ROTC unit there are two types of student—scholarship and non-scholarship. Although the focus of this book is on military programs that provide tuition aid, it should be pointed out that you may join an ROTC unit after you get to college even if you don't receive a scholarship. You take the same ROTC courses as a scholarship student, and you may major in nearly any subject. You can drop out at any time prior to the start of your junior year. If you continue you will be paid $100 per month for your last two years of college and be required to attend a summer training session between your junior and senior years. Upon graduation you will be commissioned as a second lieutenant or ensign. Your minimum active-duty obligation is three years, four years if you are in the Air Force. About 71,000 of the 95,000 students in ROTC are not on scholarship.

The major source of scholarships is the four-year tuition scholarship program. Four-year scholarships are awarded to high school seniors on the basis of a national competition. Each year approximately 3,600 winners are selected (1,000 Army, 1,300 Navy, and 1,300 Air Force) from about 25,000 applicants. Recipients of four-year Air Force and Naval ROTC scholarships may attend either a host college or cross-enrollment college. Currently the Army does not permit four-year scholarship winners to cross-enroll; they must attend a host college.

In return for the ROTC scholarship, you must serve four years on active duty, unless you choose a branch of the service (such as aviation) that requires extended training. After you accept the scholarship, you have a one-year grace period before you incur a military obligation. Prior to beginning your sophomore year, you may simply withdraw from the program. If you drop out after that time, you may be permitted to leave without penalty, ordered to active duty as an enlisted servicemember, or required to repay the financial aid you have received. The military will choose one of these three options, depending on the circumstances of your withdrawal.

Should you decide to try for a four-year ROTC scholarship, it is important that you apply to a college to which you can bring an ROTC scholarship. Because there is always the possibility you may not be accepted at your first choice, it is a good idea to apply to more than one college with an ROTC affiliation. In the case of Army and Air Force ROTC scholarships, both of which require you to major in a specified area, you also need to be admitted to the particular program for which the scholarship is offered. For example, if you win an Air Force ROTC scholarship designated for an engineering major, you must be accepted into the engineering program as well as to the college as a whole.

While the majority of new ROTC scholarships are four-year awards given to high school seniors, each service sets aside scholarships for students who are already enrolled in college and want to try for this kind of military financial aid for their last two or three years. These in-college scholarships are a rapidly growing area within the ROTC program, since the services are finding they can do a better job of selecting officer candidates after observing one or two years of college performance. Furthermore, of interest to applicants is the fact that the selection rate is quite a bit higher for the two- and three-year awards than it is for the four-year scholarship. For example, in a recent year Air Force ROTC accepted 37 percent of its candidates for four-year awards and 63 percent of its candidates for two- and three-year awards.

Most of these in-college scholarships are given to students who join an ROTC unit without a scholarship and then decide to try for a tuition grant. Since a cadet or midshipman takes the same ROTC courses whether on scholarship or not, it makes good sense for those who are not receiving aid to apply for an in-college award.

Even if you have not been a member of an ROTC unit during your first two years in college, it is possible to receive a two-year scholarship, provided you apply by the spring of your sophomore year. If you win a two-year scholarship, you will go to a military summer school where you will receive training equivalent to the first two years of ROTC courses. You then join the ROTC unit for your junior and senior years. (There are also limited opportunities for non-ROTC members to try for a three-year in-college scholarship; interested students should check with an ROTC unit.)

If you receive a two- or three-year scholarship, your active-duty obligation is four years. You will not have the one-year grace period four-year scholarship winners have in which to decide whether they want to remain in ROTC. You must make up your mind whether or not you want to stay when you attend your first military science class as a scholarship student.

You may be married and still receive an ROTC scholarship (you may not be married in the service academies). The benefits are the same regardless of whether you are married or single.

In summary, there are four ways to participate in ROTC:

1. As a winner of a four-year scholarship (or, in some cases, a three-year award) for high school seniors.

2. As a recipient of a two- or three-year scholarship for ROTC members who are not initially on scholarship.

3. By receiving an in-college scholarship (usually for two years) designated for students who have not yet joined an ROTC unit.

4. As a nonscholarship student.

The following table shows the number of students currently in various Army, Air Force, and Naval ROTC programs.

	Army ROTC	Naval ROTC	Air Force ROTC
Total number of students enrolled	60,000	10,000	25,000
Number of scholarships in effect	10,000	6,500	5,400
Four-year awards	3,000	5,300	4,000
Two- and three-year awards	7,000	1,200	1,400
Number of officers produced each year	8,000	2,000	2,700

ARMY ROTC

The evolution of the U.S. Army is intertwined with the history of the country itself—at Valley Forge with George Washington, at Gettysburg under General Meade, and in Europe and other parts of the globe during World Wars I and II. The Army, with its traditional branches of infantry, artillery, and armor, has long had the responsibility of providing land forces to defend the country in time of war. The modern Army consists of about eighteen branches, with assignments ranging from missile expert to helicopter pilot to budget officer.

The United States Military Academy, commonly called West Point, was the only source of Regular Army officers for many years after its founding in 1802. By 1900, a number of colleges offered a military training program, but it was not until the time of World War I that Congress established Army ROTC in order to create a way for young men to become Army officers directly upon graduation from civilian colleges.

Today, with Army ROTC participation possible at about 1,500 different colleges, more than 75 percent of all Army second lieutenants come from the ROTC program. In fact, Army ROTC provides about twice as many Regular (career-oriented) second lieutenants as West Point. If you are interested in an Army career, the West Point path appears to offer a small advantage over ROTC, as its graduates are more numerous among the highest-ranking officers. But overall the differences are fairly minor, and each year more and more ROTC graduates are advancing to the highest levels of the Army.

Army ROTC is the oldest and largest of the ROTC programs. The founding of Army ROTC in its present form dates to 1919 and the close of World War I, when units were established at 125 colleges across the nation. Today there are 315 colleges that host Army ROTC units and 97 extension centers of these host colleges. In addition, there are 1,082 colleges that have cross-enrollment agreements that allow a student to attend one of these colleges and commute to a host institution to participate in ROTC. A list of Army ROTC host colleges and extension centers is found in Appendix B.

Under current rules, to receive the four-year Army scholarship, you must be enrolled at one of the 315 host colleges or one of the 97 extension centers; you are not permitted to cross-enroll. You may, however, attend one of the cross-enrollment colleges and compete for a two- or three-year scholarship.

Although virtually all Army scholarship recipients attend four-year colleges, there are six military junior colleges where you can take the first two years of ROTC courses, provided you are prepared to transfer to an approved college to get your bachelor's degree.

Army ROTC offers a limited number of its four-year scholarships to high school students who want to attend a historically black college. This option is called the Quality Enrichment Program, and there are 20 participating colleges (for example, Tuskegee, Howard, and Morgan State).

As might be expected, since the Army has considerably more ROTC host units than the Navy and Air Force, as well as the most colleges with cross-enrollment agreements, it also offers the greatest number of scholarships and produces the largest number of officers. For 1989–90, there were 10,000 Army scholarships, of which approximately 1,000 were new four-year scholarships given to students entering college as freshmen. Counting cadets who are already enrolled, there are about 3,000 four-year scholarships in effect in any given year. The other 7,000 are used as in-college scholarships for either two or three years. The large scope of the Army ROTC program is important, since with an Army scholarship there is a wide range of possibilities about where to attend college and a greater number of two- and three-year scholarships for which to compete.

Army ROTC, like Air Force ROTC, uses academic quotas in awarding scholarships. In Army ROTC, 30 percent of the scholarship recipients must be engineering students. The other 70 percent can major in either the sciences or nontechnical fields.

Perhaps the most distinctive aspect of the Army ROTC program is its practice of assigning a number of its second lieutenants to the Reserve Forces, consisting of the National Guard and the Army Reserve. This means that for many Army ROTC scholarship recipients, the only active-duty requirement is to attend a six-month officers' Basic Course before joining a Reserve unit for five and a half years of service as a "weekend warrior." (Currently neither Air Force nor Naval ROTC offers this alternative—a newly commissioned officer in either of these two services must serve a minimum of four years on active duty.) If you are the type of person who would like to limit the time spent on active duty and pursue your civilian career plans, this special Reserve commission can be a very desirable option. On the other hand, if you place a high priority on being assured four years of active duty, this unique characteristic of Army ROTC may be a disadvantage. Based on its manpower needs, each year the Department of the

Army decides how many officers will be assigned to active duty and how many will go to the Reserve Forces. At present, about 50 percent of Army ROTC graduates are earmarked for the National Guard or Army Reserve.

THE FOUR-YEAR SCHOLARSHIP FOR HIGH SCHOOL SENIORS

Scholarship Benefits

Winners of the four-year scholarship receive:

1. $7000 or 80 percent of tuition, whichever is higher.

2. Payment of required fees.

3. An allowance for books.

4. $100 per month for ten months.

5. A travel allowance from home to college to begin your freshman year.

6. Payment during summer training at a rate of about $400 per month.

7. Uniforms.

8. Free flights on military passenger aircraft when there is space available.

Scholarship Obligations

In return for the scholarship, you are required to:

1. Study in the academic area in which the scholarship is offered and meet your college's requirements for a bachelor's degree.

2. Enlist in the Army Reserve, enroll in an Army ROTC unit, and complete the four-year Military Science program.

3. Complete one semester of a foreign language.

4. Attend one 6-week summer training period.

5. Upon graduation, accept a commission as an officer in either the Regular Army, the Army Reserve, or the Army National Guard.

6. Serve at least four years on active duty, unless you are assigned to the Reserve Forces, in which case your active-duty obligation can be as short as six months.

Application and Selection

Eligibility Standards

To receive a four-year scholarship, you must:

1. Be a U.S. citizen.

2. Graduate from high school but not be enrolled in college.

3. Be at least 17 years old but no more than 21 (the age limit may be extended for veterans).

4. Be accepted by a college with an Army ROTC unit on campus.

5. Plan to pursue a specific course of study.

6. Be of good moral character and have no personal convictions that prohibit you from serving in the military.

The Selection Process

You will be asked to fill out a detailed application and submit either SAT or ACT scores. Selection of finalists is based on:

1. Test scores (a minimum combined score of 850 on the SAT or a minimum composite score of 17 on the ACT is required).

2. High school academic standing.

3. Extracurricular participation, with emphasis on leadership roles and athletics.

4. Recommendations from high school officials or teachers (three are required).

Finalists are also required to:

1. Have an interview with an Army officer.

2. Pass a medical exam.

3. Pass a physical fitness test.

Intended Course of Study in College

In awarding its four-year scholarships, the Army is looking for students who possess certain academic skills. Because many of the positions in the modern Army are technical in nature, there is an emphasis on engineering and science backgrounds. Below you will see the breakdown of scholarship winners according to their intended majors in college. As you can see, the chances of being selected are highest for those who expect to study engineering, physical science, or business. They are considerably lower for those interested in social science, nursing, humanities, prelaw, or premedicine.

Engineering	30%
Physical science	25%
Business	20%
Social science	10%
Nursing	7%
Other (humanities, prelaw, premed)	8%

Medical Requirements

If you become a finalist, you must pass a comprehensive medical exam. Here are some of the important standards. You should take care of any correctable deficiencies before you report for a physical.

1. *Height and Weight*
   Weight must be proportional to height (see the chart in Appendix F).

Height Range	Men	5'0" to 6'8"
	Women	4'10" to 6'8"

2. *Eyesight*
   Distant and near vision can be correctable to 20/20, but there are limitations on muscle balance and refractive error. Color vision must include the ability to distinguish between vivid red and vivid green.

3. *Hearing*
   Allowable decibel loss varies from 25 at low frequencies to 45 at high frequencies.

4. *Allergies*
   No severe hay fever. No symptoms of asthma since age 12.

5. *Heart*
   Normal heartbeat. No hypertension or history of cardiovascular problems.

6. *Teeth*
   Numerous unfilled cavities may be a cause for disqualification.

Physical Fitness Standards

Before you can be awarded a scholarship, you must demonstrate your physical fitness by achieving satisfactory performance in the kneeling basketball throw, the standing broad jump, the 300-yard shuttle run, and pull-ups (for men) or the flexed-arm hang (for women).

Selection Timetable

Early Cycle

If you want to be a candidate for the early selection cycle, you should begin getting ready during the spring of your junior year in high school.

June	Latest date to take SAT or ACT
July 15	Application deadline
November	Army notifies winners subject to passing the medical exam

Early-cycle candidates who are not selected will be considered in the regular-cycle competition.

Regular Cycle

December	Latest date to take SAT or ACT
December 1	Application deadline
Fall/early winter	Apply to the colleges of your choice
Winter	Take the medical exam and physical fitness test

| March | Army notifies winners |
| August | Final date for medical qualification |

Keep in mind that you must be admitted to both an Army host college and the academic program you specified on your ROTC application. For example, if the Army approves your scholarship for study in engineering, your college must accept you into its engineering school.

Attending College

Your Obligation to the Army Unit

You are required to take four years of Military Science courses, including both classroom and drill sessions. In class you will learn about the Army from books; in drill periods you will learn by doing—through marching, map reading, physical fitness activities, and field exercises.

In a typical program, during the first two years you will enroll in the Basic Course for 1.5 hours of class time per week. The Basic Course consists of an introduction to the Army organization, military concepts, and management skills. In your junior and senior years, you will take the Advanced Course for 2.5 hours per week. The Advanced Course focuses on military history, tactics, military justice, and issues affecting the modern Army. The drill sessions will continue for about 2 hours a week for your entire four years. Your total time commitment will run to about 3 or 4 hours per week during the freshman and sophomore years, increasing to 4 or 5 hours per week during the junior and senior years. Depending on the college you attend, the amount of academic credit you receive for your ROTC classes will vary. On average, you will receive academic credit for about half of your Military Science courses.

In addition, once a semester there usually will be field training, an exercise in which classroom principles are tested in a military environment. You may also participate in related voluntary activities, such as a drill team or social club.

Your Academic Program in College

You will have to pass a one-semester course in a foreign language. You also will be required to continue to study in the academic area you specified when you were selected as a scholarship recipient. Any change in major will have to be approved by the Army or you will face possible loss of your scholarship. While it will not be difficult to switch majors within the same field (for example, going from chemistry to physics), you probably will have to reapply for your scholarship if you transfer to a different course of study. Scholarship students may not major in theology.

Summer Training

Between your junior and senior years, you will attend an Army Advanced Camp for six weeks. You will be there with hundreds of other cadets from colleges in your ROTC region. At summer camp you will learn about the different branches that make up the Army so you can get a better idea of the one in which you would like to serve after graduation. You also will be evaluated for your potential as a future Army officer in such areas as leadership, marksmanship, and land navigation.

Nursing students will attend a special summer training session at an Army medical facility.

Graduate School Opportunities

Although all three services prefer that newly commissioned officers proceed directly to active duty, the Army is fairly open-minded when it comes to allowing its officers to attend graduate school. Subject to Army approval, there are two ways you can do this. First, you can simply defer reporting to active duty and attend graduate school as a civilian at your own expense. Second, you can be selected for a special graduate program in which you attend college as a salaried officer with your tuition expenses paid by the Army. In this category are advanced education for the health professions and a technical enrichment program for other specialties. If you don't go to graduate school after college and you decide to stay in the Army after your first tour of duty, in all likelihood you will have additional opportunities to pursue an advanced degree.

Active-Duty Requirements

Length of Service

If you are commissioned as a Regular Army officer or a Reserve officer assigned to the active-duty Army, your minimum obligation is four years. If you are commissioned as a Reserve officer with an assignment to the National Guard or Army Reserve, you serve six months of active duty followed by seven and a half years in the Reserve Forces. As a member of a Ready Reserve unit, you attend drills one weekend per month and go on two weeks of active duty each year. During your last two years, you are placed on Standby Reserve and are not required to attend drills. If you decide to enter Army aviation, your service obligation is four years after approximately one year of flight training. Whenever the Army pays for your graduate education, your active-duty obligation is extended, usually by an additional year for each year of school.

Types of Jobs

During your senior year you will be given the opportunity to select the branch of the Army in which you would like to serve after you are commissioned as a second lieutenant. Nearly all cadets will be assigned to one of their first three choices. When you report to active duty, you will work as a supervisor and manager of resources (both people and equipment) in a position of leadership and responsibility. Second lieutenants who go to graduate school before reporting for active duty are likely to become specialists, such as physicians or lawyers.

The branches in which you may choose to serve are:

Adjutant General: personnel administration and management.

Air Defense Artillery: air defense weapons systems, including missiles, fire-control equipment, radar, and computers.

Armor: tanks and armored reconnaissance vehicles.

Aviation: helicopters and fixed-wing airplanes.

Chemical Corps: chemical, biological, and radiological activities; decontamination; and smoke production.

Corps of Engineers: Army construction projects, including production of electric power, buildings, highways, airfields, and bridges.

Field Artillery: guns, missiles, rockets, and related weapons; support equipment such as radar, range finders, and survey instruments.

Finance: pay records, fund accounting, budgets, auditing, and statistical analysis.

Infantry: rifles, mortars, antitank missiles, personnel carriers, vehicle-mounted guns, and fire-control equipment.

Medical Service: medical, dental, psychological, and social work.

Military Intelligence: information associated with military plans and operations.

Ordnance: ammunition and explosives.

Quartermaster Corps: supplies, equipment, and spare parts.

Signal Corps: radio and radar receiving and transmitting equipment.

Transportation: passenger and cargo vehicles and boats.

For more information about the four-year Army ROTC scholarship, see your high school guidance counselor, visit an Army ROTC unit on a college campus, or contact:

College Army ROTC
P.O. Box 1688
Ellicott City, Maryland 21043-9923
Telephone: 800-USA-ROTC

TWO- AND THREE-YEAR SCHOLARSHIPS FOR COLLEGE STUDENTS

If you missed out on the national competition for an Army scholarship while you were in high school, you may still try for tuition assistance if you are enrolled at any of the approximately 1,500 colleges that offer the opportunity to participate in Army ROTC. As a first step, you should visit the nearest Army ROTC unit and sign up for the Basic Course. Once enrolled, you are eligible to apply for a two- or three-year in-college scholarship.

You may apply for a scholarship for your last two or three years of college even if you were not a member of an Army ROTC unit for your freshman or sophomore year. You should realize, however, that your chances of winning a scholarship are greater if you have been a cadet, since you will have had the opportunity to show your motivation for the Army, an important factor in the selection of recipients of in-college scholarships. If you are not an Army ROTC cadet when you win a two-year in-college scholarship, you go to a special six-week Basic Camp in the summer, where you are taught the first two years of Military Science so you can join the Army ROTC Advanced Course as a junior. You may also attend the Basic Camp without a scholarship and try for one of the 300 scholarships awarded there.

If you are enrolled in a junior college, you may try for a two-year scholarship provided you are also accepted as a transfer student at an eligible four-year institution.

For in-college scholarships, the eligibility criteria, benefits, obligations, medical standards, career opportunities, and service obligations are the same as for the four-year scholarships. (See the previous pages for details.)

There are also two-year in-college scholarships for students who are majoring in nursing. Upon commissioning as second lieutenants, these graduates join the Army Nurse Corps and are assigned to one of the Army's forty-eight medical centers in the United States and overseas.

If you are a reservist who has completed two years of college, you may be eligible for a two-year Reserve Forces Duty Scholarship. Upon commissioning, you will be assigned to the Army Reserve or Army National Guard to complete an eight-year service obligation.

For more information about two- and three-year Army ROTC scholarships, contact the Professor of Military Science at an Army ROTC unit near you.

NAVAL ROTC

As a country bordered by two oceans, the United States has needed a Navy to control the seas in time of war since its first days as a republic. From the 1700s to the present, from sail to steam to nuclear power, the Navy has been a strong and effective arm of the Armed Forces. The Department includes not only the submarines, surface ships, and aircraft of the Navy, but it also embraces the Marine Corps. The Marines are the land-based extension of naval sea power, and for over 200 years they have been considered an elite branch of the U.S. military establishment.

While the Army may have had a head start on the Navy in the founding of both a service academy (West Point in 1802, Annapolis in 1845) and ROTC (Army in 1919, Navy in 1926), the Navy was the first service to establish an ROTC scholarship program that commissioned Regular officers upon graduation from civilian colleges, thus placing them on an equal footing with their colleagues from the Naval Academy. The success of the Naval ROTC (NROTC) scholarship program, called the Holloway Plan after its founder, Admiral William Holloway, cleared the way for both the Army and the Air Force to initiate their own ROTC scholarship programs.

Since their earliest days, the Naval Academy and NROTC have constituted a commissioning program for both Navy ensigns and Marine Corps second lieutenants. The relationship of the Marine Corps officer training program to the Navy program is similar to the relationship of the Marine Corps to the Navy as a whole. The Marine Corps comes under the naval organization but maintains its own distinctive identity. NROTC midshipmen who have selected the Marine option follow a Navy path for their first two years, before switching over to specific Marine training at the start of their junior year. During the academic year, future second lieutenants take courses dealing with Marine Corps subjects and take their final summer training session at Quantico, Virginia.

The Navy has the fewest ROTC host colleges of the three services, with 69 in 1989–90. In addition, there are 124 institutions that have cross-enrollment agreements, for a total of 193 colleges to choose from if you win the Naval ROTC scholarship. A list of these colleges is contained in Appendix C. Because of its relatively small size, Naval ROTC offers fewer scholarships and commissions only 2,000 Navy and Marine Corps officers each year, fewer than the Army's 8,000 and the Air Force's 2,700. For 1989–90, there were 6,500 Naval ROTC scholarships. Of the 5,300 four-year scholarships in effect in a given year, 1,300 are given to students entering college as freshmen. The remaining 1,200 awards are two- and three-year in-college scholarships used to fill vacancies that occur when four-year scholarship recipients withdraw.

Some of the three-year scholarships are given to alternates in the four-year scholarship competition. If you receive such a designation, you will join the Naval ROTC unit at your college without the benefit of financial aid for your freshman year and then be guaranteed a scholarship for three years, beginning with your sophomore year.

Though the smallest of the three ROTC programs, the Navy has the toughest medical standards and places a somewhat heavier demand on scholarship recipients while they are enrolled in college. A greater time commitment is called for in the Naval Science classes themselves, in the requirement to take other college courses, and in the summer, when an annual Navy cruise is scheduled. On the other hand, the Navy is less restrictive about your major than the Army or Air Force, since it does not have selection quotas based on academic field. Its approach is to permit you to choose from a wide range of study areas but to make sure you have adequate technical preparation by requiring math and physics courses at the college level.

THE FOUR-YEAR SCHOLARSHIP FOR HIGH SCHOOL SENIORS

Scholarship Benefits

Winners of the four-year scholarship receive:

1. Payment of tuition and academic fees.

2. An allowance for books.

3. $100 per month for ten months.

4. A travel allowance from home to college to begin your freshman year.

5. Payment during summer training at a rate of about $400 per month.

6. Uniforms.

7. Free flights on military passenger aircraft when there is space available.

Scholarship Obligations

In return for the scholarship, you are required to:

1. Study in an approved academic field and meet your college's requirements for a bachelor's degree.

2. For Navy midshipmen, complete the necessary courses in calculus, physics, English, computer science, foreign language, and national security policy; for Marine Corps option students, complete one-semester courses in military history and national security policy.

3. Enlist in the Navy or Marine Corps Reserve, enroll in an NROTC unit, and meet the Naval Science course requirements.

4. Attend three summer training periods, each four to six weeks long.

5. Upon graduation, accept a commission as an ensign in the Navy or a second lieutenant in the Marine Corps.

6. Serve a minimum of four years on active duty as part of an overall eight-year obligation.

Application and Selection

Eligibility Standards

To receive a four-year scholarship, you must:

1. Be a U.S. citizen.

2. Graduate from high school but not be enrolled in college.

3. Be at least 17 years old but no more than 21 (the age limit may be extended for veterans).

4. Be accepted by a college with an NROTC unit on campus or with a cross-enrollment agreement.

5. Plan to pursue an approved course of study.

6. Be of good moral character and have no personal convictions that prohibit you from serving in the military.

When you apply for a scholarship you will be asked to select either the Navy or Marine Corps option. You should do this based on your own sense of which service is best for you; the competition is equally strong for either program. Marine Corps training is very demanding in terms of physical fitness and discipline. You should be quite certain that you are prepared for such a rigorous regimen before you choose the Marine Corps option. Once enrolled in an NROTC unit, you may change your mind and switch from the Marine Corps to the Navy or from the Navy to the Marine Corps. Marine Corps option students make up about one sixth of the total number of Navy midshipmen.

The Selection Process

To become a Navy finalist, you need to have SAT scores of at least 430 verbal and 520 math, or ACT scores of at least 18 in English and 24 in math.

To become a Marine Corps finalist, you need to have a combined SAT score of 1000 or a combined ACT score of 45.

High school class rank and grade point average are also considered, but there are no specified minimums.

If you are chosen as a finalist, you are next required to:

1. Complete an application that includes your high school transcript, a report of your extracurricular activities, and teacher recommendations.

2. Have an interview with a panel of Navy officers.

3. Take a medical exam.

Intended Course of Study in College

Unlike the Army and Air Force, the Navy and Marine Corps do not have selection quotas based on academic field. You may major in virtually any subject except the health professions (such as medicine, dentistry, and veterinary science).

Medical Requirements

Applicants must pass a medical exam to qualify for an NROTC scholarship; in addition, Marine Corps candidates must pass a physical fitness test. The standards below do not constitute a complete list but are intended to give you a general idea of the medical requirements. You should take care of any correctable deficiencies before you report for a physical.

On the day of your medical exam, your blood will be tested for the presence of drugs, alcohol, and the AIDS virus. A positive AIDS test is disqualifying.

1. *Height*

Height Range	Men	Navy	5'2" to 6'6"
		Marines	5'6" to 6'6"
	Women	Navy	5'0" to 6'6"
		Marines	5'0" to 6'6"

2. *Weight*

Weight must be proportional to height, with the following body-fat limits (see the chart in Appendix F):

Men	Navy	26%
	Marines	18%
Women	Navy	36%
	Marines	26%

3. *Eyesight*

   Distant and near vision: 20/20 in each eye without correction. Up to one third of scholarship recipients may receive waivers if their vision is correctable to 20/20, provided refractive error does not exceed ±5.50 diopters. Normal color vision is required for the Navy. There is no color vision requirement for the Marine Corps.

4. *Hearing*

   Allowable decibel loss varies from 30 at low frequencies to 60 at high frequencies.

5. *Allergies*

   No severe hay fever. No symptoms of asthma since age 12.

6. *Heart*

   Normal heartbeat. No hypertension or history of cardiovascular problems.

7. *Teeth*

   Must be in excellent dental health.

Selection Timetable

The Naval ROTC national scholarship competition operates on a rolling schedule, with selection boards meeting monthly, beginning in October. The earlier you complete all the application requirements, the earlier you will learn if you have won the scholarship.

November	Latest date to take SAT or ACT
Fall/early winter	Apply to the colleges of your choice
Midwinter	Take the medical exam and have an interview; Marine Corps candidates also take the physical fitness test
Late January	Application deadline
October to April	Navy notifies winners

Attending College

Your Obligation to the Navy Unit

You will be required to take four years of Naval Science courses. Classes meet for 3 hours each week, and the drill period takes an additional 1.5 hours per week. In the freshman year, the courses are Introduction to Naval Science and Naval Engineering; for the sophomore year, Naval Weapons and Seapower; for the junior year, Navigation and Naval Operations; and for the senior year,

Leadership and Management. If you have chosen the Marine Corps option, you will take Evolution of Warfare and Amphibious Warfare in place of the Navy courses in your junior and senior years. During drills you will participate in training that extends beyond the Naval Science courses, specifically military formations, physical conditioning, and practical exercises.

At a typical college, you will receive academic credit for nearly all your Naval Science courses, which total from 22 to 24 credit hours. Therefore, although NROTC classes may be more time-consuming than those offered by the Army and Air Force, it is somewhat more likely that these courses will count as part of your normal academic work load.

The NROTC unit will also sponsor activities in addition to classes and drills. As a midshipman, you will have the opportunity to participate in athletic and social functions or be a member of a drill team.

Your Academic Program in College

NROTC midshipmen are expected to take one-year courses in calculus, physics, English, and national security policy, as well as one-semester courses in computer science and a foreign language. Marine Corps option students are required to take one-semester courses in military history and national security policy.

Summer Training

You will have military duty each summer. As a Navy midshipman, you will serve aboard a ship during the summer prior to your sophomore year and again in the summer before your senior year. During the summer before your junior year, you will be introduced to amphibious warfare and flight training. Marine Corps option students will join their Navy classmates for the first two summers before attending camp at Quantico, Virginia, prior to their senior year.

Graduate School Opportunities

Although the Navy or Marine Corps sometimes permits a student to delay reporting to active duty to attend graduate school, the vast majority of new officers proceed directly to Navy duty stations or to the Marine Corps Basic School. Opportunities for graduate school occur later in the career of a Navy or Marine Corps officer.

Active-Duty Requirements

Length of Service

The minimum obligation is four years for both the Navy and Marine Corps. If you choose submarine duty, you are required to serve four years after approximately one and a half years of training. If you decide on either Navy or Marine Corps aviation, your minimum active duty is five years after flight training, which runs from one to one and a half years.

Types of Jobs

Navy

After you are commissioned as an ensign, you can serve in one of three branches of the Navy.

1. Surface Warfare. Here, after a brief period of specialized training, you will be assigned to sea duty on board a ship. Regardless of the type of ship— tanker, destroyer, or aircraft carrier—your duties will involve management, leadership, and knowledge of naval operations.

2. Submarines. Before reporting to a submarine as a division officer, you will first undergo a rigorous graduate program in nuclear power and propulsion systems followed by specific training aboard a submarine. You will then proceed to the fleet.

3. Aviation. You may be either a flight officer or a pilot. In either case, you will be assigned to flight training for twelve to eighteen months before reporting to a fleet squadron. Naval flight officers serve as crew members and are in charge of navigation and weapons systems. Naval pilots fly a variety of aircraft, many of which operate from aircraft carriers.

Marine Corps

After you are commissioned as a second lieutenant, you will attend the Basic School at Quantico, Virginia, for twenty-one weeks of further training. In Basic School you will continue your Marine Corps professional education, particularly in such areas as marksmanship, infantry tactics, and amphibious operations. Based on the knowledge and training you gain in Basic School, you will select one of the Marine Corps specialties: Infantry, Tracked Vehicles, Aviation (Marine Corps pilots go through Navy flight training), Field Artillery, Engineering, Communications, Supply, and Computer Science. Unless you are selected for flight training, you will proceed to the Fleet Marine Force after Basic School.

For more information about the four-year Naval ROTC scholarship, see your high school guidance counselor, visit the local Navy recruiting office, check with a Naval ROTC unit on a college campus, or contact:

NROTC Scholarships
P.O. Box 27429 NW
Washington, D.C. 20038
Telephone: 800-327-NAVY

TWO- AND THREE-YEAR SCHOLARSHIPS FOR COLLEGE STUDENTS

As with the Army and Air Force, you may enroll in an NROTC unit as a nonscholarship student and take the same Navy courses as scholarship recipients. If you then wish to apply for a scholarship for the remaining two or three years of your undergraduate education, you request a nomination from the Professor of Naval Science and enter the competition for an in-college scholarship. If you decide to continue as a nonscholarship student, you will have a three-year service obligation after being commissioned in the Navy or Marine Corps Reserve.

It is also possible to receive a two-year Navy scholarship if you are not a member of an NROTC unit. If you are interested, you should contact the Professor of Naval Science in the fall of your sophomore year. If you are selected, you attend the Naval Science Institute during the summer and then join the NROTC unit for your junior and senior years. You should bear in mind, however, that your chances of receiving a scholarship are better if you are already a member of an NROTC unit, since you will have a military "track record" that is helpful in the selection process.

For in-college scholarships, the eligibility criteria, benefits, obligations, medical standards, career opportunities, and service obligation are the same as for the four-year scholarships. (See the previous pages for details.)

To find out more about two- and three-year scholarships, check with a Naval ROTC unit near you.

AIR FORCE ROTC

The Air Force is the youngest of the military services. In the early 1900s, as military air power began to grow, the Army created a special division called the Army Air Corps. At the conclusion of World War II, Congress decided that the use of aircraft in warfare had become so important that the United States needed a military branch specializing in air power. As a result, Congress passed a law in 1947 that created the U.S. Air Force as a separate service.

Air Force ROTC (AFROTC) was also established in 1947 and has continued to commission officers for forty-three years. In 1955, the Air Force Academy was founded, providing another way for the Air Force to commission second lieutenants directly upon their graduation from college.

The Air Force has the same mixture of scholarship programs for college students as the other services—the four-year national scholarship competition for high school students, various in-college scholarships for undergraduates, and enrollment in the Air Force Academy. There are, however, distinctive characteristics of the Air Force's programs that are important for you to consider as you think about which service is right for you.

Within the Air Force as a whole, the Air Force Academy is virtually the only source of Regular officers, as well as the primary producer of pilots and navigators. So if you are fairly sure you want to become a career Air Force officer or you want to enter flight training, you should try first for the Air Force Academy. The Air Force ROTC program, by contrast, has a separate role: to supply Reserve officers who will serve on active duty as scientists or engineers in areas that support Air Force flight operations. There is a clear distinction between the role of Academy graduates and that of ROTC graduates. While there are some exceptions, the general rule is that the Air Force Academy produces Regular officers who will be assigned to flight duties, while the AFROTC produces Reserve officers who will provide scientific, engineering, and management support.

This is quite different from the arrangement in the Navy (and the Army, to some extent), where both academy and ROTC graduates are commissioned as Regular officers and are assigned to the same kinds of active-duty jobs. The more limited role of an Air Force second lieutenant who graduates from ROTC will appeal to some people for various reasons. For example, the physical standards are more lenient for AFROTC than for the other scholarship programs. Also, it is likely that your active-duty assignment will be closely related to your academic field. If, for example, you major in aeronautical engineering, you probably will work with avionics systems rather than in personnel or supply.

There are 150 colleges that host Air Force ROTC units and another 455 institutions that have cross-enrollment agreements. A list of these colleges is contained in Appendix D. A cross-enrollment arrangement means you can attend a college near another college that has an AFROTC unit on campus and commute to take the ROTC courses. A student attending any one of the four-year colleges among the 605 may hold an Air Force scholarship—either the four-year award for high school seniors or an in-college scholarship for students already enrolled. If you are attending a two-year college, you need to be accepted as a transfer student by a four-year college in order to qualify for the scholarship.

For 1989–90, there were 5,400 AFROTC scholarships, of which approximately 4,000 were for four years, and 1,400 were two- and three-year in-college scholarships.

THE FOUR-YEAR SCHOLARSHIP FOR HIGH SCHOOL SENIORS

Scholarship Benefits

Winners of four-year scholarships receive:

1. Payment of tuition and fees. The AFROTC divides its tuition scholarships into two categories. Type I awards, which go to 20 percent of recipients, provide full tuition. Type II awards, which go to the remainder of recipients, pay tuition up to a cap of $7500. Eighty percent of colleges with AFROTC units have tuition under $7500; however, if you choose to attend a college with a higher tuition, you will have to pay the difference. It is possible for Type II recipients who perform well during their first two years in AFROTC to apply for a Type I scholarship.

2. An allowance for books.

3. $100 per month for ten months.

4. A travel allowance from home to college to begin your freshman year.

5. Payment during summer field training at a rate of about $400 per month.

6. Uniforms.

7. Free flights on military passenger aircraft when there is space available.

If you apply for a four-year AFROTC scholarship and your credentials are not quite as strong as those of the primary winners, you may still receive a scholarship for three and a half or three years. The application process and benefits received are the same as for a four-year scholarship, except that tuition payments do not start until the second semester of your freshman year or the beginning of your sophomore year.

Scholarship Obligations

In return for the scholarship, you are required to:

1. Study in the academic area in which the scholarship is offered and meet your college's requirements for a bachelor's degree.

2. Enlist in the Air Force Reserve, enroll in an AFROTC unit, and complete the four-year Aerospace Studies program.

3. Complete one year of a foreign language.

4. Attend one 4-week summer training period.

5. Upon graduation, accept a commission as a second lieutenant in the Air Force Reserve.

6. Serve at least four years on active duty.

Application and Selection

Eligibility Standards

To receive a four-year scholarship, you must:

1. Be a U.S. citizen.

2. Graduate from high school but not be enrolled in college (you may be a part-time college student and still be eligible for a scholarship; check with an AFROTC unit).

3. Be at least 17 years old but no more than 21 (the age limit may be extended for veterans).

4. Be accepted by a college with an AFROTC unit on campus or one with a cross-enrollment agreement.

5. Plan to pursue a specific course of study.

6. Be of good moral character and have no personal convictions that prohibit you from serving in the military.

The Selection Process

To become a finalist, you need to:

1. Have a high school grade point average of at least 2.5 on a 4.0 scale.

2. Be in the top 25 percent of your high school class.

3. Have a combined SAT score of at least 1000 (minimum 500 math, 450 verbal) or a composite ACT score of at least 23 (minimum 20 math, 19 English).

If you are chosen as a finalist, you are required to:

1. Complete an application that includes your high school transcript, a report of your extracurricular activities, and teacher recommendations.

2. Have an interview with a panel of Air Force officers.

3. Take a medical exam.

Intended Course of Study in College

Your chances of winning a four-year scholarship vary according to the subject you plan to major in. The Air Force allocates ROTC scholarships to fill its need for officers with certain skills. Although the Air Force requirements vary somewhat from year to year, there is a strong emphasis on engineering majors, with some openings for scientists. Very few scholarships are given to liberal arts majors. For 1990–91, the scholarships are divided as follows:

Engineers	80%
Science majors (architecture, computer science, math, meteorology, and physics)	18%
Nontechnical majors (accounting, business, economics, and management)	2%

Unless you are fairly certain you want to major in engineering or science, AFROTC is probably not the program for you, as the Army and Navy have considerably more openings for liberal arts majors.

Medical Requirements

If you become a finalist, you must pass a comprehensive medical exam. The standards shown below are not meant to be comprehensive but are intended to give you a general idea of the medical requirements. You should take care of any correctable deficiencies before you report for a physical.

1. *Height and Weight*
   Weight must be proportional to height (see the chart in Appendix F).
   Height Range   Men   5'0" to 6'8"
                       Women   5'0" to 6'8"

2. *Eyesight*
   Correctable to 20/20. Refractive error in each eye cannot exceed ±8.00 diopters. The standards for color vision vary depending on Air Force specialty.

3. *Hearing*

Allowable decibel loss varies from 25 at low frequencies to 90 at high frequencies.

4. *Allergies*

No severe hay fever. No symptoms of asthma since age 12.

5. *Heart*

Normal heartbeat. No hypertension or history of cardiovascular problems.

6. *Teeth*

Numerous unfilled cavities may be a cause for disqualification.

Selection Timetable

The Air Force chooses scholarship winners on a rolling basis, with selection boards meeting in November, January, and March. The earlier you apply, the greater your chance of acceptance.

December	Latest date to take SAT or ACT
December 1	Application deadline
Fall/early winter	Apply to the colleges of your choice
Late fall/winter	Take the medical exam
December, February, and March	Air Force notifies winners

Keep in mind that you must be admitted to both an Air Force host college and the academic major you specified on your ROTC application. For example, if the Air Force approves your scholarship for study in engineering, your college must accept you into its engineering school.

Attending College

Your Obligation to the Air Force Unit

You are required to take four years of AFROTC courses, called Aerospace Studies. During your first two years in college, you take the General Military Course for 2 hours per week—1 hour of class and 1 hour of drill (called Leadership Laboratory). The General Military Course covers an introduction to the military and the development of U.S. air power. In the junior and senior years, the time commitment increases to 4 hours per week—3 hours of class in the Professional Officer's Course and 1 hour of drill. In the Professional Officer's Course you learn about management skills and American defense policy. During the drill periods you participate in military formations, physical fitness training, and leadership exercises.

Depending on the terms of the agreement between the ROTC unit and the college, you may receive academic credit for your AFROTC courses. At a typical college, about 75 percent of the Aerospace Studies courses count toward your graduation requirement. If an AFROTC course is not approved for credit, you will have to take the course in addition to your normal academic work load.

Besides taking part in AFROTC classes and drills, you will have the opportunity to participate in voluntary activities sponsored by the unit. These usually include a drill team, social events, and trips to Air Force bases.

Keep in mind that if you are interested in flying, either as a pilot or as a navigator, the AFROTC scholarship program is probably not the path you should be taking. If you realize this after you've enrolled in AFROTC, you may be able to qualify for navigator school and have your scholarship shifted to that duty assignment. However, if you are selected for pilot training, there is a good chance that you will lose the last two years of your scholarship, since the Air Force has a surplus of pilot candidates and therefore does not normally pay tuition costs for such students.

Your Academic Program in College

Besides taking specific AFROTC courses, you must make satisfactory progress in the major subject the Air Force has approved and complete a year of a foreign language. The Air Force is firm about requiring you to stay with the academic field you specified when you were selected. If you switch your major to a subject that is in a different academic area (for example, from engineering to science, or from science to liberal arts), you will lose your scholarship but be given the chance to compete for a scholarship in the new field. If you are successful, you will regain your financial aid. You should realize, however, that the scholarship competition is very tough if you try to move from engineering to either science or liberal arts.

Summer Training

You are required to attend one 4-week field training camp at an Air Force base, normally in the summer between your sophomore and junior years. This is your opportunity to learn firsthand what the Air Force is really like and to think seriously about the type of active-duty assignment you would like upon graduation.

Graduate School Opportunities

You may request a delay in reporting to active duty to attend graduate school at your own expense. You may also apply for a program in which the Air Force pays for your graduate education while you serve on active duty. The Air Force offers a number of such opportunities in areas ranging from health professions to meteorology. The largest of these is the Minuteman Education Program—a special program for missile launch officers that can lead to a graduate degree in business or management.

If you don't go on to graduate school directly after college and you decide to stay in the Air Force after your first tour of duty, in all likelihood you will have the opportunity to continue your education at a later time. The Air Force places a strong emphasis on graduate study, and 60 to 70 percent of its officers eventually obtain advanced degrees.

Active-Duty Requirements

Length of Service

If you are not involved in flying, you must serve a minimum of four years on active duty. Navigators have a five-year obligation after a training period of approximately six months. Pilots serve six years after about a year of flight training. If the Air Force pays for your graduate education, your active duty is extended in proportion to the time it takes you to get an advanced degree. There are exceptions to this rule, however. For example, the Minuteman Education Program is designed to enable you to obtain an advanced degree within your normal service obligation.

Types of Jobs

Upon graduation you will be commissioned as a second lieutenant in the Air Force Reserve and ordered to active duty. The duty assignment you receive is highly dependent on the technical skills you acquired while you were in college and your overall academic and military record. Flight personnel and missile officers come from a variety of academic backgrounds, but, as with most Air Force jobs, a good background in math and science is important. Beyond these two assignment areas, there is a fairly close correlation between your academic major and the type of active duty you can expect. Here are some examples:

College Major	Air Force Duty
Administration/Management	Accounting and Finance, Personnel, Supply, Computer Systems, Food Services, Space Systems
Arts, Humanities, or Education	Air Intelligence, Education and Training, Public Affairs
Computer Science	Computer Systems, Computer Operations
Engineering	Space Systems Analyst, Aircraft and Missile Maintenance, Project Engineer, Research, Management Analysis, Manufacturing Engineer
Law	Judge Advocate
Medical Science	Physician, Dentist, Pharmacist, Therapist, Nurse
Science	Space Systems, Communications, Weather, Research, Computer Systems, Scientific Analyst, Air Intelligence
Social Studies	Air Intelligence, Education and Training, Management, History, Executive Support

For more information about four-year Air Force ROTC scholarships, see your high school guidance counselor, visit an AFROTC unit on a college campus, or talk to an Air Force Regional Admissions Counselor. You can find out the location of an Air Force counselor near you by contacting:

AFROTC
Maxwell Air Force Base
Alabama 36112-6663
Telephone: 205-293-2091

TWO- AND THREE-YEAR SCHOLARSHIPS FOR COLLEGE STUDENTS

Nationwide, there are about 800 new AFROTC scholarships awarded to college students each year, either to cadets who were not previously on scholarship or to undergraduates who have not yet joined a unit. Both two- and three-year scholarships are available to students who have sufficient time remaining in their academic programs (including graduate work) to meet the scholarship requirements. For example, if you applied for a three-year scholarship as a sophomore, you would have to be enrolled in a five-year academic program, such as engineering or architecture. Like the four-year scholarship for high school students, most of the in-college scholarships are awarded to science or engineering majors. There are, however, two-year scholarships available for students in nontechnical fields, as well as for navigators and missile officers.

Although in-college scholarships are open to all students, your chances of selection are improved if you are a member of an AFROTC unit. In practice, virtually all the three-year scholarships go to AFROTC cadets. Therefore, if you do not sign up as a freshman, the two-year scholarship is normally the only one for which you will be able to compete. Furthermore, if you are already a cadet you have the opportunity to show your motivation for the military service, an important plus in the selection process. A student who wins a scholarship but who is not already a member of an AFROTC unit is required to take the equivalent of the first two years of Aerospace Studies at a special summer session. If you are a junior college student, you have to be accepted at a four-year institution before you can receive a two- or three-year scholarship.

The eligibility criteria, benefits, obligations, medical standards, active-duty opportunities, and service obligation for AFROTC in-college scholarships are the same as for the four-year scholarship for high school seniors. (See the previous pages for details.)

The Air Force has special in-college scholarships for students majoring in the health professions. There are two- and three-year AFROTC scholarships for undergraduates who are pursuing a premedical degree. After graduation, recipients of these scholarships attend medical school while serving on active duty. Two-year scholarships also are available to students who are majoring in nursing. Upon graduation, Air Force ROTC nurses enter an internship program as their first duty assignment.

If you would like more information about an in-college scholarship, check with a college AFROTC unit near you.

ONE-YEAR SCHOLARSHIPS FOR LAW, METEOROLOGY, AND NURSING MAJORS

Students within one year of graduation who are majoring in meteorology or nursing or attending law school are eligible for a one-year scholarship. The benefits are full tuition and a stipend of $100 per month. The in-college obligation is six weeks of field training during the summer prior to enrollment and completion of Aerospace Studies 300—Leadership and Management in your final year of college; after graduation, you must complete the summer course, Aerospace Studies 400—National Security Forces. Upon completion of the latter, you are commissioned and assigned to a position in your field of expertise. Your service obligation is four years. The AFROTC unit at your college has further information about this program.

SUMMARY AND COMPARISON OF ROTC SCHOLARSHIP PROGRAMS

The following table summarizes the important aspects of the ROTC scholarship programs.

	Army	Navy/ Marine Corps	Air Force
Eligibility Standards	**Four-year scholarships:** U.S. citizen, high school graduate, age 17–21, plan to pursue approved course of study, of good moral character, no personal convictions against serving in the military. **Two- and three-year scholarships:** all of the above, plus enrolled in college, maintaining satisfactory progress toward degree.		
Medical Requirements			
Eyesight Normal vision	Correctable to 20/20.	20/20 without correction, with some waivers.	Correctable to 20/20.

(continued)

Summary and Comparison of ROTC Scholarship Programs (continued)

	Army	Navy/ Marine Corps	Air Force
Color vision	Distinguish between vivid red and vivid green.	Navy: normal; Marines: no requirement.	Requirement varies by specialty.
Height Men	5'0" to 6'8"	Navy: 5'2" to 6'6"; Marines: 5'6" to 6'6"	5'0" to 6'8"
Women	4'10" to 6'8"	5'0" to 6'6"	5'0" to 6'8"

Scholarship Benefits	Full or partial tuition, books, $100 per month for ten months, travel allowance, pay for summer training, uniforms, free flights on military aircraft when space permits.
Military Requirements	Enlist in the Reserves, complete the ROTC military science program (may hold scholarship for one year before obligation begins). **Two- and three-year scholarships:** obligation commences upon enrolling in military science course.

	Army	Navy/Marine Corps	Air Force
Academic Requirements	Take one or two semesters of a foreign language, meet the college's requirements for a bachelor's degree.		
	Study in a specified academic field.	Navy: take the necessary courses in calculus, physics, English, national security policy, and computer science; Marines: take one-semester courses in military history and national security policy.	Study in a specified academic field.

	Army	Navy/Marine Corps	Air Force
Summer Training	One summer.	Three summers.	One summer.
	Two- and three-year scholarships: usually one or two summers, depending on previous ROTC affiliation.		

	Army	Navy/ Marine Corps	Air Force
Type of Commission	Regular or Reserve second lieutenant.	Regular Navy ensign or Marine second lieutenant.	Reserve second lieutenant (a few commissioned as Regulars).
Length of Active Duty	Unless assigned to the Reserves, minimum of four years.	Minimum of four years.	Minimum of four years; five years for navigators; six years for pilots.
Where You Can Enroll	315 host colleges or 97 extension centers. **Two- and three-year scholarships:** 1,082 cross-enroll-ment colleges.	194 host or cross-enrollment colleges.	607 host or cross-enrollment colleges.
Academic Quotas	30% engineering 25% physical science 20% business 10% social science 7% nursing 8% other **Two- and three-year scholarships:** Army and Air Force quotas may be somewhat different, depending on needs of the service.	None	80% engineering 18% science 2% other
Selection Calendar			
Early	SAT/ACT by June. Apply in July. Notification in November.	None.	None.
Regular	SAT/ACT by December.	Rolling acceptance.	SAT/ACT by December.

(continued)

Summary and Comparison of ROTC Scholarship Programs (continued)

Army	Navy/ Marine Corps	Air Force
Apply by December. Notification in March.	SAT/ACT by November. Apply October to January. Notification from October to April.	Apply by December. Notification in December, February, and March.

Two- and three-year scholarships: application due fall or early spring of freshman or sophomore year at ROTC unit, decisions made in late spring or summer.

Chapter 3

The Service Academies

Next to ROTC, the most common way to combine college with training as a military officer is by attending a service academy. The Army, Navy, Air Force, Coast Guard, and Merchant Marine have their own academies. Marine Corps officers are trained at the Naval Academy. These academies offer a technically oriented education equal to that at any of the top engineering schools in the country, with all college expenses paid by the federal government.

If you go to college at a service academy, you will be a full-time member of the military and, except for vacation periods, you will wear a uniform and be subject to military rules and regulations. The life-style is a regimented one, and a strict disciplinary code is in force. However, the financial rewards for this more extensive military involvement are proportionately greater than they are if you choose the ROTC path.

Before you decide to apply to a service academy, you should have a strong feeling that the military is an occupation that appeals to you. You should not apply merely because an academy offers a free education or because someone else thinks it is the place for you. You should be fairly settled on the idea of a career as a military officer. This does not mean you have to be 100 percent certain; that is too much to expect of a 17- or 18-year-old who is only beginning to think seriously about future plans. On the other hand, your career goals should be at least roughly defined and you should be able to picture yourself serving as an officer for a number of years.

The military academies offer a very rigorous and special type of education. Your academic interests should be in pursuing a Bachelor of Science rather than a liberal arts degree. On the personal side, you should be the type of person who is willing to become completely involved in military life as a college student. You will attend an academy for four years, with only short periods off for vacations. College life will consist of a well-structured daily schedule, a heavy academic work load, adherence to an honor code, and considerable physical activity.

The nature of cadet or midshipman life at a service academy can be illustrated by describing a typical yearly calendar and the daily routine.

Yearly Calendar

Freshman Summer Indoctrination	July 1–August 15
Fall Semester	August 16–December 22
Christmas Vacation	December 23–January 8
Winter Semester	January 9–March 10
Spring Vacation	March 11–17
Spring Semester	March 18–May 16
Summer Vacation	May 17–June 17
Summer Camp/Cruise	June 18–August 15

In other words, as a cadet or midshipman at a service academy, out of a full year you will attend class for thirty-six weeks, have summer training for eight weeks, and be on vacation for eight weeks.

Daily Routine

6 a.m.	Wake up
6:45–7:15 a.m.	Breakfast
7:15–7:30 a.m.	Morning formation
7:30 a.m.–noon	Morning classes
Noon–1 p.m.	Lunch
1–3:30 p.m.	Afternoon classes
3:30–6 p.m.	Military drill, athletics, extracurricular activities
6–7 p.m.	Dinner
7–11 p.m.	Study
11 p.m.	Lights-out

This schedule usually applies five and a half days a week, from Monday morning to Saturday noon. Upperclassmen normally have liberty from Saturday noon until Sunday night. Time off for freshmen is quite restricted, with only an occasional opportunity to leave the academy grounds. The amount of liberty and other privileges varies with seniority (the higher your class, the more free time you get), responsibility, and performance.

Although the number of officers produced by the service academies each year is less than the number that comes through the ROTC programs, the academies are a very important source of officers, particularly among those who intend to make the military their career. The following table summarizes the sizes of the different academies.

	West Point	Annapolis	Air Force	Coast Guard	Merchant Marine
Number of Cadets or Midshipmen	4,500	4,500	4,500	900	1,000
Number of Officers Produced Each Year	1,000	1,000	950	150	250

THE ADMISSIONS PROCESS AT WEST POINT, ANNAPOLIS, AND THE AIR FORCE ACADEMY

Because of the similarities in the process by which West Point, Annapolis, and the Air Force Academy offer appointments (the military term for admission), the descriptions for all three are combined in the following. Because the Coast Guard Academy and Merchant Marine Academy follow a somewhat different admissions procedure, their guidelines are given separately.

Eligibility Standards

You need to be a U.S. citizen, unmarried (and continue to be so while enrolled at an academy), of good moral character, and at least 17 but no more than 22 years old in the year of admission.

Precandidate Questionnaire

First, you will fill out a brief, two-page questionnaire that registers you as an applicant. If you meet the basic eligibility standards, the academy will send you a package of admissions material that includes information about how to ask for a nomination.

Requesting a Nomination

The next step is to be nominated by an official source, normally your congressman. Each congressman has a set number of nominees he or she can recommend for admission. The military service itself may also nominate candidates who are sons and daughters of past and present members of the Armed Forces. The process of obtaining a nomination is not as complicated as it appears: you simply write standard letters to the representative from your district and to your senators. You do not have to have political influence, nor is it necessary that you know your congressman. The nomination procedure is merely an additional screening that takes place within your state or congressional district.

Getting Admitted

If you receive a nomination, you go through the next phase of the admissions process, which involves evaluation of your SAT or ACT scores, a review of your high school record including teachers' recommendations, a physical fitness test, and a medical exam. The academy admissions office offers appointments to the most qualified among those who have met all the standards. It looks carefully at a combination of factors, weighing academic performance, high school activities, and strength of character. West Point, Annapolis, and the Air Force Academy all have rolling admissions plans. Since they accept students as applications are completed, it is to your advantage to apply early.

Chapter 1 contains information about the kinds of high school courses you should take if you are seriously considering a service academy. It also contains the profile of a typical successful applicant. You may wish to reread that section.

COSTS

At the four military academies—West Point, Annapolis, the Air Force Academy, and the Coast Guard Academy—the cost of tuition, room, and board is paid by the federal government. Cadets also receive about $500 per month, from which the academy deducts the cost of uniforms, books, and laundry. During your freshman year you will have approximately $60 remaining each month to cover personal expenses. From your sophomore year on you will have more spending money available, since most of your uniform costs are charged in your first year. There is also a travel allowance based on the distance from your home to the academy. You will be asked to make a deposit of approximately $1000 to meet initial expenses until your paychecks come in. Although it offers a nearly free education, the Merchant Marine Academy has a somewhat different arrangement. See the appropriate section.

THE UNITED STATES MILITARY ACADEMY

Located in West Point, New York, the United States Military Academy prepares young men and women for careers as Army officers. West Point is quite selective in its admissions. Each year approximately 14,000 students apply and 5,800 receive nominations. Of these, 2,900 are judged to be fully qualified scholastically, medically, and physically. About 1,900 are eventually admitted, and about 1,400 accept admission into the freshman class.

Following is a summary of things you should know about West Point.

Admissions Calendar

Spring of junior year in high school (preferred) or early senior year	Fill out precandidate questionnaire
Spring of junior year (preferred) or early senior year	Request a nomination

Early Selection

No later than November	Take SAT or ACT
By December 1	Complete application
By January	The Army notifies those who receive an appointment

Regular Selection

No later than February	Take SAT or ACT
By March 21	Complete application
During the West Point admissions processing period	Take medical exam and physical fitness test
November to April	The Army notifies those who receive an appointment

Medical Requirements

The standards are the same as for the Army ROTC scholarship; see Chapter 2.

Physical Fitness Requirements

To be offered an appointment, you have to pass the West Point physical aptitude exam. This test consists of a shuttle run, kneeling basketball throw, standing long jump, and pull-ups (for men) or flexed-arm hang (for women).

Graduation Requirements

You will be expected to:

1. Attend West Point for four years—fall and spring semesters and a summer session each year.

2. Major in one of four fields of study: applied sciences and engineering, basic sciences, humanities, or national security and public affairs.

3. Maintain a grade point average (GPA) of at least 2.0 (C).

4. Successfully pass all required courses and physical and military training; show satisfactory conduct, including adherence to the honor code.

5. Graduate with a Bachelor of Science degree and accept a Regular Army commission as a second lieutenant.

Active-Duty Requirements

Service Obligation

The minimum service requirement is five years. This time commitment will be extended if you attend graduate school at the Army's expense. Your service obligation may also increase if you attend certain Army branch schools, including flight training.

Types of Duty

Most newly commissioned second lieutenants report to their Army duty station soon after graduation, although a number of outstanding students win scholarships or fellowships for graduate study. Even if you don't go directly to graduate school from West Point, there will be many opportunities to pursue an advanced degree later in your career.

A West Point graduate will enter one of the branches of the Army listed in the Army ROTC section in Chapter 2 and be expected to exercise important leadership and management responsibilities as a second lieutenant. By law, at least 80 percent of cadets must be commissioned in a combat unit.

For more information about West Point, see your high school guidance counselor or contact:

Director of Admissions
United States Military Academy
West Point, New York 10996-1967
Telephone: 914-938-4041

Your high school guidance counselor should also be able to put you in touch with the West Point Field Representative in your area.

THE UNITED STATES NAVAL ACADEMY

The United States Naval Academy is located in Annapolis, Maryland, on the shore of the Chesapeake Bay, about an equal distance from Baltimore and Washington, D.C. Annapolis looks for young men and women who are highly motivated toward either the Navy or the Marine Corps.

Since Annapolis attracts a large number of well-qualified applicants, it is selective in its admissions. In a typical year there are about 15,000 applications. Of these, 6,600 receive nominations and 2,500 are judged to meet the Academy's scholastic, physical, and medical standards. About 1,700 are admitted and 1,350 decide to attend.

Admissions Calendar

Spring of junior year in high school (preferred) or early senior year	Fill out precandidate questionnaire
Spring of junior year (preferred) or early senior year	Request a nomination
January 31	Last date to submit a nomination

No later than February	Take SAT or ACT
	Complete application
During the Annapolis admissions processing period	Take medical exam and physical fitness test
October to April	The Navy notifies those who receive an appointment

Medical Requirements

The standards are the same as for the four-year NROTC scholarship, as described in Chapter 2.

Physical Fitness Requirements

You must pass a physical fitness test consisting of pull-ups (for men) or the flexed-arm hang (for women), standing broad jump, kneeling basketball throw, and 300-yard shuttle run.

Graduation Requirements

You will be expected to:

1. Attend Annapolis for four years—fall and spring semesters and a summer session each year.

2. Major in engineering, science, or math, unless you have special approval to major in the humanities or social sciences.

3. Maintain a GPA of at least 2.0 (C).

4. Successfully pass all required courses and physical and military training; show satisfactory conduct, including adherence to the honor code.

5. Graduate with a Bachelor of Science degree and accept a commission as an ensign in the Regular Navy or a second lieutenant in the Regular Marine Corps.

Active-Duty Requirements

Service Obligation

Your active-duty obligation is a minimum of five years. Should you choose submarines or aviation, your length of service will be extended by the twelve to eighteen months it takes to go through training.

Types of Duty

Since both Annapolis and NROTC graduates are commissioned as Regular officers in the Navy or Marine Corps, you will be assigned to the types of jobs that are listed in the NROTC section in Chapter 2. As with the policy for ensigns and second lieutenants who come through NROTC, there are only a few opportunities for graduate study right after college, and these are primarily for outstanding students who have been successful in a fellowship competition. Even if you do not go to graduate school directly from Annapolis, there will be many opportunities to pursue an advanced degree later in your career.

For more information about Annapolis, see your high school guidance counselor or contact:

Director of Candidate Guidance
United States Naval Academy
Annapolis, Maryland 21402-5018
Telephone: 800-638-9156

Your guidance counselor should also be able to give you the name of the Naval Academy Information Officer in your area.

THE UNITED STATES AIR FORCE ACADEMY

Located in Colorado Springs, Colorado, the United States Air Force Academy is the source of nearly all Regular Air Force officers, 75 percent of whom enter flight training as either pilots or navigators. To consider applying to the Air Force Academy, you should be seriously interested in a career in the Air Force, most likely as a pilot or navigator.

The Air Force Academy is highly selective. In a typical year its admissions office receives 14,000 applications. Of these, 6,400 are nominated, 3,500 meet the academic, physical, and medical standards, and 1,850 are admitted. On average, 1,400 of the admitted candidates decide to enroll.

Admissions Calendar

Spring of junior year in high school (preferred) or early senior year	Fill out precandidate questionnaire
Spring of junior year (preferred) or early senior year	Request a nomination
No later than the end of November	Take SAT or ACT
November	Apply for early selection
By end of January	Apply for regular selection
During the Academy admissions processing period	Take medical exam and physical fitness test
November to April	The Air Force notifies those who receive an appointment

Medical Requirements

Air Force Academy nonflying personnel must meet the same medical standards as Air Force ROTC scholarship recipients (see Chapter 2). Those who plan to enter flight training must meet the following eyesight requirements and height restrictions.

Pilot: Uncorrected 20/20 vision. Normal color vision. Height 5'0" to 6'8".

Navigator: Distant vision correctable to 20/20 but no worse than 20/70 uncorrected. Near vision correctable to 20/20 in one eye and 20/30 in the other. Normal color vision. Height 5'0" to 6'8".

Physical Fitness Requirements

You must pass a physical fitness test consisting of push-ups, sit-ups, pull-ups, and a 300-yard shuttle run.

Graduation Requirements

You will be expected to:

1. Attend the Air Force Academy for four years—fall and spring semesters and a summer term each year.

2. Major in any one of twenty-three subject areas, including science, engineering, social science, and humanities.

3. Maintain a GPA of at least 2.0 (C).

4. Successfully pass all required courses and physical and military training; show satisfactory conduct, including adherence to the honor code.

5. Graduate with a Bachelor of Science degree and accept a Regular Air Force commission as a second lieutenant.

Active-Duty Requirements

Service Obligation

For nonflying officers, the minimum service obligation is five years. Navigators serve five years after about six months of training. Pilots are required to sign up for six years after approximately one year of flight training.

Types of Duty

Most newly commissioned second lieutenants go directly on active duty, although a number of outstanding students win scholarships or fellowships for graduate study. Even if you do not proceed to graduate school right away, there will be many opportunities to pursue an advanced degree later in your Air Force career.

If you are not involved in aviation, you will be assigned to the types of active-duty jobs described in the AFROTC section in Chapter 2.

If you are one of the 75 percent earmarked for flight training, you will end up as an aviator in either the Tactical Air Command (fighter aircraft), the Strategic

Air Command (bombers), or a support group like the Military Airlift Command (transport planes).

For more information, see your high school guidance counselor or contact:

Cadet Admissions Office
United States Air Force Academy
Colorado Springs, Colorado 80840
Telephone: 719-472-2640

Your guidance counselor should also be able to give you the name of the Air Force Liaison Officer in your area.

THE COAST GUARD ACADEMY

The Coast Guard is one of the oldest of the military services, founded in 1790 when President George Washington asked Congress for a number of ships to guard the shores of the young republic. The U.S. Navy was formed eight years later as the seagoing military force, but the need for a special service to patrol U.S. coastal waters remained and, in one form or another, has continued to the present day. During peacetime, the Coast Guard comes under the Department of Transportation and carries out eight primary duties: (1) encouraging boat safety; (2) conducting search and rescue missions; (3) maintaining aids to navigation; (4) regulating the Merchant Marine; (5) dealing with marine environmental protection; (6) enforcing maritime law; (7) ensuring port safety; and (8) conducting ice patrols.

When the United States is engaged in war, the Coast Guard becomes part of the Navy and may be called to serve anywhere in the world. For example, during World War II Coast Guard cutters operated out of Greenland to help convoy American ships, and in the Vietnam War the Coast Guard was used in Southeast Asia to prevent the movement of enemy shipping.

The Coast Guard does not sponsor an ROTC program, so the only way you can go through officer training while in college is by attending the U.S. Coast Guard Academy. The Academy, founded in 1876, is located in New London, Connecticut, on the banks of the Thames River, not far from where the river flows into Long Island Sound. Like West Point, Annapolis, and the Air Force Academy, the Coast Guard Academy is highly selective in its admissions and looks for top-notch students who have strong personal qualities and are interested in serving their country as military officers. The most distinctive aspect of the Coast Guard Academy's admissions process is that it works like that of any other competitive college. Applicants submit the required material directly to the Coast Guard Academy admissions office, and the most qualified are offered admission. Unlike applicants to the other four academies, a Coast Guard Academy applicant does not have to be nominated by a congressman. The absence of this requirement makes the Coast Guard Academy's admissions process more straightforward and easier to deal with.

During a recent year, 6,900 students applied, and 520 were admitted. Of the admitted students, 280 actually enrolled. The Coast Guard Academy is considerably smaller than West Point, Annapolis, or the Air Force Academy, with 900 cadets compared to 4,500 at each of the other three. It is about the same size as the Merchant Marine Academy.

The main difference between the Coast Guard and the other military services lies not so much in the nature of the Academy, but rather in the mission of the Coast Guard itself. As parts of the Department of Defense, the purpose of the Army, Navy, Air Force, and Marine Corps is to defend the United States against foreign aggression. In contrast, the Coast Guard, under the Department of Transportation, spends much of its time in a nonmilitary role—making waterways safer for boating, easier to navigate, and freer of accidents, oil spills, and other dangers. If this part of the Coast Guard's work—saving lives and protecting property—appeals to you, the Coast Guard Academy bears a close look.

Although the type of work you will do after you become an officer in the Coast Guard may be quite different from the jobs you would hold in the other services, student life at the academies is similar. (See the yearly calendar and daily routine at the beginning of this chapter.) In your studies, you will concentrate on science or engineering, with some courses in other areas. Outside of the classroom, you will be completely immersed in a military life-style, with a well-structured daily schedule, adherence to discipline and an honor code, and considerable physical activity. You will either attend classes or be involved in summer cruises for all but about two months of the year.

High School Academic Preparation

The Coast Guard Academy has a minimum course requirement of 3 years of math and 3 years of English. A fourth year of both math and English is recommended, along with courses in American history, a laboratory science, and a foreign language.

Eligibility Standards

You must be a U.S. citizen, unmarried (and continue to be so while enrolled at the Academy), of good moral character, and between 17 and 22 years old in the year of admission.

Getting Admitted

After you complete the admission application, your credentials will be evaluated on the basis of your high school grades, scores on either the SAT or ACT, teachers' recommendations, leadership potential as demonstrated by extracurricular activities, sports and community involvement, and interest in becoming a Coast Guard officer. The strength of a candidate's academic credentials is the most important single factor, counting for 60 percent in the admissions decision, but the Academy also looks at motivation, leadership, and personal outlook.

These attributes account for the remaining 40 percent of the admissions formula.

The profile of a typical Coast Guard Academy cadet is similar to that of students at the other academies, as described in Chapter 1.

Admissions Calendar

June to December	Submit application
September through December	Take SAT or ACT
By January 15	All application material due
During the Coast Guard Academy admissions processing period	Take medical exam
February 15	The Coast Guard notifies finalists
Early April	Accepted students are notified

Medical Requirements

To be admitted to the Coast Guard Academy, you must meet the medical standards for a commissioned officer. The standards shown below are not meant to be comprehensive but are intended to give you a general idea of the medical requirements. You should take care of any correctable deficiencies before you report for a physical.

1. *Height and Weight*
   Weight must be proportional to height (see the chart in Appendix F).
   Height Range (men and women)   5'0" to 6'6"

2. *Eyesight*
   Distant and near vision correctable to 20/20, with uncorrectable vision no worse than 20/200. Refractive error may not exceed ±5.50 diopters. Normal color vision is required.

3. *Hearing*
   Allowable decibel loss varies from 30 at low frequencies to 55 at high frequencies.

4. *Allergies*
   No severe hay fever. No symptoms of asthma since age 12.

5. *Heart*
   Normal heartbeat. No hypertension or history of cardiovascular problems.

6. *Teeth*
   Numerous unfilled cavities may be a cause for disqualification.

Physical Fitness Requirements

By the end of your first summer, you will have to pass a physical exam that consists of pull-ups, sit-ups, standing long jump, shuttle run, and 1.5-mile run.

Graduation Requirements

You will be expected to:

1. Attend the Coast Guard Academy for four years—fall and spring semesters and a training period each summer.

2. Major in one of these fields of study: civil engineering, electrical engineering, marine engineering, marine science, mathematics and computer science, government, or management (75 percent of cadets major in a technical area; the remaining 25 percent major in either government or management).

3. Maintain a GPA of at least 2.0 (C).

4. Successfully pass all required courses and physical and military training; show satisfactory conduct, including adherence to the honor code.

5. Graduate with a Bachelor of Science degree and accept a Regular commission as an ensign in the U.S. Coast Guard.

Active-Duty Requirements

Service Obligation

The minimum service obligation is five years.

Types of Duty

Upon being commissioned as an ensign, graduates proceed to active duty. The majority of assignments are on board a Coast Guard ship, for example, on a cutter conducting search and rescue, an icebreaker in the polar regions, or a buoy tender maintaining navigation aids.

Because Coast Guard ships carry relatively small crews, officers take on major responsibilities early in their career. Before your first five-year tour of sea duty is completed, it is likely that you will be one of the higher-ranking officers on board ship, perhaps even the commanding officer of a small vessel.

The Coast Guard also has an aviation branch you may apply for after one year of sea duty. If selected, you will attend flight training under direction of the Navy. After you receive your wings, you will fly Coast Guard patrol aircraft. Your total service obligation will be extended by the time it takes you to complete flight training.

Although there are no graduate school opportunities directly after you receive your commission, applications are accepted after you serve aboard ship for one year. If you go to graduate school, your active-duty time will increase by one year for each year of graduate school.

For more information about the Coast Guard Academy, see your high school guidance counselor or contact:

Director of Admissions
United States Coast Guard Academy
New London, Connecticut 06320-4195
Telephone: 203-444-8501

THE MERCHANT MARINE ACADEMY

Although the Merchant Marine is a civilian service—a privately owned fleet of ships that transport goods all over the world—it is also part of the U.S. defense framework. This is because merchant vessels are responsible for importing the strategic materials that are necessary for America's defense, as well as for delivering military supplies overseas. The Merchant Marine is known as the "fourth arm of defense" and works closely with the Navy in time of war.

The Merchant Marine is best remembered for its efforts during World War II. At that time the American commercial shipping industry was given the responsibility for transporting thousands of soldiers and millions of pounds of military supplies to the U.S. Armed Forces in Europe and throughout the Pacific. It was largely through the work of the Merchant Marine that the United States was able to extend its power across two oceans at the same time.

Beyond the seagoing duties of the Merchant Marine, there is an extensive shore establishment that supports the shipping industry. It includes ports and terminals, shipyards, admiralty lawyers, and engineering and research companies.

Like the Coast Guard, the Merchant Marine does not train officers through an ROTC program. Therefore, if you want to graduate from college as a licensed mate or engineer, fully trained and ready to assume duties aboard a commercial ship, the U.S. Merchant Marine Academy at Kings Point, New York, on the North Shore of Long Island, may be the place for you. The primary difference between Kings Point and the service academies is that the other four train officers for the military, while Kings Point produces officers who will work in a civilian occupation.

In spite of the nonmilitary nature of the Merchant Marine, the Navy maintains close ties with Kings Point. Each student at Kings Point is appointed as a midshipman in the Naval Reserve and takes a Naval Science curriculum. Furthermore, graduates of Kings Point are commissioned as officers in the Naval Reserve so they can keep abreast of the current state of affairs in the Navy while serving in their civilian occupation.

As a midshipman at the Merchant Marine Academy, your college experience will be quite similar to that at any of the military academies. You will have a heavy academic load that emphasizes science and engineering, and your non-classroom activities will be structured along regimental lines. You will be under the leadership of active-duty Navy officers, and you will be required to wear a uniform, adhere to discipline, and follow the honor code.

The Merchant Marine Academy is also like the military service academies in its admissions procedures. It is highly selective, seeking well-rounded people who combine a strong academic background with characteristics of self-discipline and good citizenship. In a typical year the Admissions Office at Kings Point receives 2,500 applications. Of the 1,800 who receive nominations, 600 are accepted and about 300 decide to enroll. With a total student body of 1,000 midshipmen, the Merchant Marine Academy is considerably smaller than West

Point, Annapolis, or the Air Force Academy and is about the same size as the Coast Guard Academy.

Costs

The costs of tuition, room, and board are paid by the federal government. There is a fee of approximately $850 for other expenses. You are not paid while you are at the Merchant Marine Academy, but when you take the shipboard part of your academic training you receive an income of approximately $500 per month.

High School Academic Preparation

The Merchant Marine Academy has the following minimum academic requirements for its applicants: 3 years of English, 3 years of mathematics, and 1 year of a laboratory science, either physics or chemistry. In addition, a fourth year of math, both physics and chemistry, and courses in mechanical drawing and machine shop are desirable.

Eligibility Standards

You must be a U.S. citizen, of good moral character, and between 17 and 25 years old in the year of admission. You may be married, although midshipmen must reside on campus, apart from their spouse.

Preliminary Screening

You should take the courses mentioned above, rank in the top 40 percent of your class, and have a minimum combined score of 950 on the SAT or 46 on the math and English sections of the ACT.

Requesting a Nomination

If you meet the minimum qualifications, you can ask for a nomination. Nominations are made by the congressmen in your home state. During the spring of your junior year in high school, you should write to your representative and senators to request a nomination. It is not necessary for you to know your congressmen, nor do you need to have any political connections. Senators and representatives evaluate the credentials of candidates from their area and make recommendations to the Kings Point Admissions Office.

Getting Admitted

When you request a nomination, you should also submit to Kings Point the admission application, including your high school transcript, a record of your extracurricular activities, results of the SAT or ACT, and letters of recommendation. The Kings Point Admissions Office will select the best qualified from among those who receive nominations and meet the minimum academic and medical standards. Like the other services, the Merchant Marine Academy looks first at your academic record but also weighs your leadership potential and strength of character.

Admissions Calendar

Spring of junior year in high school (preferred) or early senior year	Fill out precandidate questionnaire
Spring of junior year (preferred) or early senior year	Request a nomination
No later than February	Take SAT or ACT
By March 1	Apply
During the Kings Point admissions processing period	Take medical exam
March and April	Kings Point notifies those who receive an appointment

Medical Requirements

To be admitted to the Merchant Marine Academy, you must meet the medical standards for a commission in the Naval Reserve. The standards shown below are not meant to be comprehensive but are intended to give you a general idea of the medical requirements. You should take care of any correctable deficiencies before you report for a physical.

1. *Height and Weight*
   Weight must be proportional to height (see the chart in Appendix F).
   Height range   Men   5'2" to 6'6"
                   Women   5'0" to 6'6"

2. *Eyesight*
   Distant and near vision correctable to 20/20, with uncorrectable vision no worse than 20/200. Refractive error may not exceed ±5.50 diopters. Normal color vision is required.

3. *Hearing*
   Allowable decibel loss varies from 30 at low frequencies to 60 at high frequencies.

4. *Allergies*
   No severe hay fever. No symptoms of asthma since age 12.

5. *Heart*
   Normal heartbeat. No hypertension or history of cardiovascular problems.

6. *Teeth*
   Numerous unfilled cavities may be a cause for disqualification.

Graduation Requirements

You will be expected to:

1. Attend Kings Point for four years—fall and spring semesters and a summer training period each year.

2. Major in one of four areas: marine transportation, marine engineering, dual license (a combination of marine transportation and engineering), or marine engineering systems.

3. Maintain a GPA of at least 2.0 (C).

4. Participate in two half-year periods at sea on U.S. merchant vessels.

5. Successfully pass all required courses (including Naval Science), the U.S. Coast Guard license exam, all required certifications, and physical and military training; show satisfactory conduct, including adherence to the honor code.

6. Graduate with a Bachelor of Science degree and a merchant license and accept a commission in the U.S. Naval Reserve.

After Graduation

As a licensed officer in the Merchant Marine, you are required to serve as an employee in the maritime industry of the United States for at least five years, either on board ship or in another area approved by the Secretary of Transportation. Your first shipboard position will be as Third Mate or Third Assistant Engineer. Your pay and benefits will depend on the contract you sign with your employer. Keep in mind that you also will be a member of the Naval Reserve while you go about your civilian occupation. As such, you will be required to attend drills every month and serve two weeks of active duty each year.

Because a Merchant Marine Academy midshipman receives the same kind of military training as a student at Annapolis, a limited number of graduates may choose to become full-time Navy officers and fulfill their five-year service obligation in that manner.

Special Note for Women

The Merchant Marine Academy was the first academy to become coeducational, accepting women in 1974. Since the Merchant Marine is not a military service, there are no restrictions on the type of duty to which a woman may be assigned. As Naval Reserve officers, women are subject to the limitations on assignments to combat positions that have been mentioned previously. However, as a practical matter, Reserve officers are usually far removed from combat, and women should encounter very few restrictions in military duty.

For more information, see your high school guidance counselor or contact:

Admissions Office
United States Merchant Marine Academy
Kings Point, New York 11024-1699
Telephone: 516-773-5000 (ask for Admissions)

Your guidance counselor should be able to tell you who the Kings Point Information Representative is for your area.

THE SERVICE ACADEMIES COMPARED

The following table summarizes the important characteristics of the military academies.

	West Point	Annapolis	Air Force	Coast Guard	Merchant Marine
Type of Academy	————————Military.————————				Civilian, but close connection with the Navy.
Eligibility Standards	U.S. citizen, high school graduate, age 17–22, good moral character, no personal convictions against serving in the military, unmarried.				The same, except age 17–25 and may be married.
Academic Requirements	————None.————			3 years math, 3 years English.	3 years math, 3 years English, 1 year lab science.
Medical Requirements	Same as for AROTC scholarship.	Same as for NROTC scholarship.	Same as for AFROTC scholarship, except flight training has its own height and vision standards.	Similar to those for NROTC scholarship.	Similar to those for NROTC scholarship, except eyesight standards are more lenient.
Scholarship Benefits	Free tuition, room, and board. Pay of $500 per month, travel allowance, free flights on military aircraft.				Free tuition, room, and board. Pay for shipboard training. An $850 charge for other expenses.

	West Point	Annapolis	Air Force	Coast Guard	Merchant Marine
Military Requirements	Enlist in the Reserves, pass military training, adhere to standards of discipline. May resign prior to end of sophomore year without penalty.				
Graduation Requirements	Attend four years, major in approved subject, maintain at least 2.0 average, pass required courses and physical training, comply with honor code, graduate with Bachelor of Science degree.				
Summer Training	——————— Attend four summer sessions. ———————				
Type of Commission	Regular second lieutenant.	Regular Navy ensign or Marine second lieutenant.	Regular second lieutenant.	Regular ensign.	Licensed as Merchant Marine officer; commissioned ensign, Naval Reserve.
Length of Active Duty	——————— Minimum of five years. ———————				
Admissions Calendar					
Preliminary Application	———Spring of junior year——— or early senior year.			None.	None.
Nomination	———Spring of junior year——— or early senior year.			None.	Spring of junior year or early senior year.
Application Deadline	March.	February.	January.	January.	March.
Notification	——————— Fall through spring of senior year. ———————				
Pay and Benefits	———Standard military pay and benefits.———				Based on contract with employer.

Chapter 4

Special Programs

In addition to ROTC and the service academies, there are a number of other programs in which the military will pay you money for college in return for your promise to become an officer. None of these options requires that you participate in military activities during the academic year. Specific officer training takes place either during the summer or after you graduate from college.

NAVY NUCLEAR PROPULSION OFFICER CANDIDATE PROGRAM

This is a path to a commission for engineering and science majors who want to serve in the nuclear Navy. It is highly selective, but those who are chosen receive substantial financial benefits.

You may apply for the Nuclear Propulsion Officer Candidate Program as a sophomore or junior in college. The Navy is looking for math, physics, chemistry, or engineering majors with high grade point averages. If accepted, you will be paid a $4000 bonus and a minimum of $1200 each month until you graduate. You will also be eligible for other military benefits, such as free medical care and Navy exchange privileges, while you are in college.

There are no military requirements while you are in college. You must maintain good grades and graduate on time. After receiving your degree, you attend the Navy Officer Candidate School for four months. Upon commissioning, you incur a five-year active-duty obligation that begins with six months at nuclear power school, followed by six months at a nuclear "prototype" (a land-based model of a shipboard propulsion system) and attendance at either a submarine school or surface warfare school before reporting to the fleet.

For more information, contact a Navy Recruiting Station.

NAVY CIVIL ENGINEERING CORPS COLLEGIATE PROGRAM

Under this program, juniors and seniors enrolled in engineering or architecture receive $1200 a month while they are in college, to a maximum of $28,000. After graduation, they attend Officer Candidate School and, upon completing the course, are commissioned as ensigns. They then receive advanced training at the Civil Engineering Corps school in California. There is a four-year service obligation.

For more information, contact a Navy Recruiting Station.

NAVY BACCALAUREATE DEGREE COMPLETION PROGRAM

If you are already attending college and have no plans to study in the technical areas required for the Nuclear Propulsion or Civil Engineering Corps options, this program may be of interest to you. It places no restrictions on your college major, and you receive $1200 each month you are enrolled. (There is a special baccalaureate degree completion program for students majoring in nursing.) While you are a student, you will have no military requirements to meet; you need only maintain a good academic record. Upon graduation from college, you will report to Officer Candidate School. Your basic active-duty obligation is four years, but this may be extended depending upon the branch of the Navy you select.

For more information, contact a Navy Recruiting Station.

MARINE CORPS PLATOON LEADERS CLASS

This program is the largest single source of Marine Corps officers and provides a way for you to become a second lieutenant without a time commitment during the academic year. You may apply during any of your first three years in college. As a freshman or sophomore, you will attend two 6-week training sessions at Quantico, Virginia. If you are a junior, there is one 10-week session. You are entitled to receive $100 a month for up to three years, a maximum of $2700 in pay.

A unique aspect of the Platoon Leaders Class (PLC) is that the time you spend in this program counts toward "longevity" in determining your salary once you become a second lieutenant. For example, if you sign up for the PLC as a college freshman, you will enter the Marine Corps as a second lieutenant with three years of qualifying service. This means you will receive $4000 more in pay than the officers who come in through other programs.

Upon being commissioned as a second lieutenant, you will first go to Basic School and then either directly to the Fleet Marine Force or, if you have been selected for aviation, to flight training. There is a special option in the PLC program for students pursuing a law degree. For more information, contact a Marine Corps Officer Selection Office.

ARMED FORCES HEALTH PROFESSIONS SCHOLARSHIP PROGRAM

The Army, Navy, and Air Force offer 5,000 scholarships to students who are attending accredited civilian medical schools in the United States or Puerto Rico. You apply for this program during your senior year in college at the same time that you are admitted to medical school. The benefits are tuition and fees, books, and a stipend of approximately $600 per month. Upon matriculation you are commissioned as a second lieutenant or ensign. During your first summer vacation period you receive basic officer training. Each summer thereafter you serve on military active duty for forty-five days. The length of your required service depends on how many years you receive the scholarship; the typical obligation is four years.

In the Air Force section of Chapter 2, there is a description of two- and three-year pre–health profession scholarships for undergraduates enrolled in ROTC. Students who successfully complete this program and are accepted to medical school automatically qualify for the Health Professions Scholarship Program.

If you would like to know more about this program, contact the Medical Officer Recruiting Office for the service you are interested in. Overall coordination for all three branches takes place in the Office of the Assistant Secretary of Defense (Health Affairs), The Pentagon, Washington, D.C. 20301.

UNIFORMED SERVICES UNIVERSITY OF THE HEALTH SCIENCES

This medical education program is similar to the Armed Forces Health Professions Scholarship Program. The main difference is that, instead of attending a civilian medical school, you enroll at the Uniformed Services University of the Health Sciences in Bethesda, Maryland. You apply as you would to any other medical school. There is no charge for tuition, since your education is subsidized by the federal government. All students are paid, active-duty military officers and are normally graduates of ROTC, an academy, or Officer Candidate School. You may also apply as a civilian and be trained as an officer while you are enrolled. Graduates have a seven-year service obligation.

For more information, contact:

Director of Admissions
Uniformed Services University of the Health Sciences
National Naval Medical Center
4301 Jones Bridge Road
Bethesda, Maryland 20014
Telephone: 301-295-3101

AIR FORCE COLLEGE SENIOR ENGINEERING PROGRAM

This is a small program for engineering majors who are in their junior or senior year in college. Approximately 75 students are selected each year.

If you are selected, you enlist as an Airman First Class and are paid at that grade (a rate of about $725 per month) until you receive your bachelor's degree. You then go to Officer Training School for about three months. Upon completion of training, you are commissioned as a second lieutenant in the Air Force and begin a four-year service obligation. Your duty assignment will be in the engineering field in which you majored.

Chapter 5

Weighing the Options

Up to this point, *How the Military Will Help You Pay for College* has described the various officer training programs that provide financial aid to help you pay for college. What you have read so far has been a compilation of facts—eligibility standards, scholarship obligations, how to apply, what to expect in college, and the types of jobs available to you after graduation. To complete the picture, what you now need are some "rules of thumb," general guidelines that will help you figure out which path would be most suitable for you.

CONSIDERING THE ACADEMIES

If you want to be a career military officer, first think about a service academy. Except for relatively minor differences, your undergraduate experience at any of the service academies will be much the same. You will wear a uniform, lead a structured life, go to college virtually all year round, receive a very good technically oriented education, and graduate with a Bachelor of Science degree, all at the government's expense. After graduation you will enter the service well prepared and well motivated to embark on a career. When thinking about which of the academies you want to try for, concentrate mainly on how you see yourself fitting in with the type of postcollege duty and life-style that a particular service offers.

The Merchant Marine Academy is distinctive, since nearly all graduates become civilian officers either on privately owned ships engaged in commerce or as part of the maritime shore establishment. As an employee of the shipping industry, you are not subject to military regulations, and your salary is based on a contract with your employer rather than on the military pay schedule. Even though you are trained to pursue a civilian occupation, there nevertheless is a close connection to the Navy. Upon graduation you will be commissioned as an officer in the Naval Reserve. Although nearly all Merchant Marine Academy ensigns are assigned to the Ready Reserve and fulfill their service obligation as "weekend warriors," you also have the option of joining the active-duty Navy and becoming a full-time military officer.

The Coast Guard Academy can also be distinguished fairly easily. Although the Coast Guard is classified as a military service, it is part of the Department of Transportation, and its primary mission is to save lives and protect property. It does carry out militarylike functions, usually during wartime, when it comes under the Navy, but also in peacetime, when it enforces maritime law. As a Coast Guard officer you probably will work from a base in the United States, be assigned to a small ship, and assume considerable responsibility at an early age.

The three large academies—West Point, Annapolis, and the Air Force Academy—are quite similar in a number of respects. Their admissions standards are about the same, they are of comparable size, the quality of education is equally strong, graduates are commissioned as Regular officers in a military service that is part of the Department of Defense, and active-duty jobs are quite varied, with considerable time spent outside the United States. Because of these similarities, you should base your choice among the Army, Navy, Air Force, or Marine Corps on how your abilities and aspirations match the type of duty you can expect after graduation rather than on less important differences in location, life-style, and academic programs. To do this most effectively, you should reread the sections on types of duty in Chapters 2 and 3, refer to the academies' own publications, and talk with individuals who have served in the military.

CONSIDERING THE ROTC SCHOLARSHIPS

If you are willing to consider a career as a military officer but aren't quite sure, and you prefer a normal college experience to life at an academy, try for the four-year ROTC scholarship. Choosing the right ROTC program can be even more complicated than deciding among the academies, since there are more differences among ROTC programs than among the academies. If you have a strong reason for wanting to be in a particular service, go ahead and try for that ROTC scholarship and plan an academic program that will give you the best chance for selection. If, however, you have no clear preference for the Army, Navy, Air Force, or Marine Corps, you should turn to secondary considerations and ask yourself the following questions.

1. *What is my area of academic interest?*
   If you want to be an engineer you will be attractive to all four services, but the Air Force selects a significantly higher proportion of engineers than the Army or Navy. If you plan to be a liberal arts major, your odds of being selected are better with the Navy and Marine Corps, somewhat lower with the Army, and lowest with the Air Force.

2. *How will I do on the medical exam?*
   The main difference among the services is in the vision standards. The Army and Air Force allow less than normal vision, provided it is correctable to 20/20. The Navy and Marine Corps, except for a limited number of waivers, require 20/20 eyesight. You can see what other differences

there are in the medical standards by rereading those sections in Chapter 2.

3. *How much time do I want to devote to ROTC during the school year and summer?*

The Army and the Air Force require less of a time commitment than the Navy and Marine Corps. The Army and Air Force Military Science courses total approximately 16 credit hours, and the only other course requirement is one or two semesters of a foreign language. There is one summer training period. Four years of Naval ROTC courses add up to 22–24 credit hours. In addition to a foreign language requirement, Navy midshipmen take calculus, physics, English, computer science, and national security policy, and Marine Corps option students take military history and national security policy. There are three summer training periods.

4. *How closely do I want my military duty to relate to my academic major?*

The AFROTC scholarship program is designed to attract engineering majors who will have technical jobs after graduation. The Army, Navy, and Marine Corps are much more open-minded about the field you major in. They are looking for well-rounded, well-educated officers who will use their academic background in exercising general leadership and management skills.

5. *Do I want to do graduate work before I go on active duty?*

Your chances of going to graduate school prior to entering active duty are higher with the Army and Air Force and lower with the Navy and Marine Corps.

6. *How important is a Regular commission to me?*

Air Force ROTC commissions nearly all its second lieutenants as Reserve officers. Therefore, an Air Force ROTC graduate starts out on a less advantageous career path than a Regular officer from the Air Force Academy.

The Navy and Marine Corps commission four-year scholarship students as Regular officers, the same designation that Naval Academy graduates receive. This places Naval ROTC graduates on an equal footing with ensigns and second lieutenants from Annapolis.

The Army has three commissioning alternatives. All West Point graduates and some ROTC graduates go into the Regular Army. Other ROTC graduates are designated Reserve officers and are assigned to four years of active duty, while still others are commissioned as second lieutenants with the National Guard or Army Reserve and go to a six-month Basic Course before becoming "weekend warriors." Army ROTC scholarship recipients who fall into this last category have the shortest active-duty service requirement of all ROTC scholarship recipients.

There is the possibility that you may decide the four-year ROTC scholarship program is for you but you are not accepted. In this case, it would be a good idea to join an ROTC unit as a freshman in college and try for an in-college scholarship as soon as one becomes available. This is a particularly worthwhile strategy if you were classified as an "alternate" when you applied for the four-year ROTC scholarship, because your chance of winning a three-year scholarship is very good.

DEFERRING THE ROTC DECISION

If you aren't very interested in the military as a high school student, wait until you get to college to make a decision about joining ROTC. Then you can visit an ROTC unit on campus and decide if you want to sign on as a nonscholarship student. Should you decide to join, stay with it for at least a semester to see how you get along. If you remain a member of the unit, then think seriously about applying for an in-college scholarship.

OFFICER TRAINING PROGRAMS FOR THOSE ALREADY IN COLLEGE

If you're already in college and the idea of serving as a military officer looks attractive, either visit an ROTC unit on your campus and see what it has to offer or think about joining one of the special officer training programs. As long as you apply by the spring of your sophomore year, you should be eligible for a two-year scholarship program. You make up the first two years of ROTC training by attending a summer training camp prior to your junior year. You then join an ROTC unit for your last two years and receive a commission upon graduation. Whether you choose the Army, Navy, Air Force, or Marine Corps depends not only on your personal preference but also on which service offers a program for you. For example, both Army and Air Force ROTC have two-year scholarships for nursing majors, and the Air Force has a program for premedical students.

The special programs described in Chapter 4 are all non-ROTC ways a college student can start training to become an officer and receive financial aid at the same time. You can find out more about these programs at the appropriate recruiting station.

PART II

Going into the Military First: College Money for Enlisted Servicemembers

The first part of *How the Military Will Help You Pay for College* described officer training opportunities available to college students who enroll in college directly after high school with the intention of becoming an officer and how the money provided (either free tuition or direct pay) can be important in helping finance a college education. However, you can also get educational benefits from the military by enlisting in the Army, Navy, Air Force, Marine Corps, or Coast Guard right after high school. When you sign on as a private, seaman, or airman, you become eligible for two large sources of aid that will reduce the cost of taking college courses and getting your college degree.

The first category consists of programs you can participate in while you are on active duty, so-called "in-service" education. These range from earning college credit for your military specialty (for example, a radar technician receiving credit for completing military electronics school as well as for on-the-job training) to taking courses at a local college with a 75 percent reduction in tuition.

The second type of college aid available to enlisted servicemembers is "after-service" education, part of the New G.I. Bill that went into effect on July 1, 1985. Under the New G.I. Bill, the military puts up about $8 for every dollar you contribute toward your college education. This arrangement enables you to build a college savings account that you can draw on after you leave the service.

Besides these two sources of college money that are open to all enlisted servicemembers, there is aid available if you want to move from the enlisted ranks to become an officer. This can be done through ROTC, a service academy, or a special program that leads to a commission.

The five military services are very supportive of continuing education programs as a means of improving servicemembers' job skills and preparing them for employment in civilian life. In a typical year, military members take approximately 700,000 courses through in-service education programs. It is estimated that the New G.I. Bill is providing in excess of $500-million in yearly benefits.

It should be pointed out that Part II focuses on the ways in which the military helps enlisted men and women work toward a college degree. Graduate school opportunities available to officers are mentioned in Part I. The Tuition Assistance program explained in Chapter 7 also is available to officers. Service academy graduates and ROTC scholarship recipients are not eligible for assistance under the New G.I. Bill. Officers who enter the military through other routes, such as the special programs and Officer Candidate School, can receive New G.I. Bill benefits under the same rules described in Chapter 8 for enlisted servicemembers.

If you have further questions about the information presented here, you should talk with a military recruiter. Recruiters can describe the options that are available and can look at your credentials and give you advice on which programs fit your qualifications.

Chapter 6

Enlisting in the Military—Is It for You?

For many young people in high school, going on to college right away is not a path that appeals to them. Instead, they prefer to take a break from school by going to work—either in a regular civilian job or in one of the military services. The Armed Forces know that many young men and women weigh this decision each year, and they rely on their recruiting centers to point out the advantages of serving in the military as a "first job" after high school graduation. This is a large undertaking, since the military is the largest employer of high school graduates in the United States, "hiring" about 300,000 each year.

There are, of course, many reasons why you might want to join the military, for example, serving your country, learning a job skill, or traveling to different parts of the world. This part of the book describes one of them—the opportunity to begin your college education part-time while you are on active duty and then pay for college when you get out by using the New G.I. Bill.

As the first step in trying to decide whether the military might be a good first job for you, you should learn more about what it is like to serve as an enlisted man or woman. Although there are some major differences in the missions of the military services, overall the services have a great deal in common when it comes to recruiting practices, basic training, types of duty, promotions, and levels of pay. This chapter describes what you can expect if you enlist in the Armed Forces.

JOINING THE SERVICE

The five military services—Army, Navy, Air Force, Marine Corps, and Coast Guard—are looking for the same type of person: a man or woman aged 17 to 35 who is a U.S. citizen or permanent resident, in good physical condition, of good moral character, and motivated to serve his or her country. A high school diploma, although not an absolute requirement, is highly recommended. (Currently about 90 percent of enlistees have graduated from high school.) All enlistees must also meet minimum standards on a military aptitude test.

After deciding which military service you want to enter, you report to a processing center where you take a medical exam and a series of aptitude tests. These results, combined with your level of education, determine whether you receive a regular assignment or qualify for a special program that includes such features as guaranteed job selection, an enlistment bonus, and accelerated promotion.

When you join the Armed Forces you sign an enlistment contract. The military agrees to provide a job and a salary. In return, you agree to serve for a certain period of time. The contract is for eight years, divided between active duty and Reserve duty. A typical enlistment is for three to six years, although the Army offers a two-year option. After your active-duty obligation ends, you spend the rest of the eight years in the Reserves. It is also possible to join the Reserve Forces directly; this is discussed later in the chapter.

TRAINING

Whichever service you join, you begin with basic training. Typically, this initial exposure to the military consists of ten weeks of intensive training during which you engage in physical conditioning, field exercises, and drills and ceremonies. You also learn about military regulations, the use of weapons, and the organization of the military. Beyond general military training, you will be taught specific subjects that are related to the service in which you enlist, such as the practice of seamanship in the Navy and Coast Guard. Basic training consists of a rigorous daily routine that begins at 6 a.m. and continues until 10 p.m. The schedule is somewhat lighter on weekends, but there is little free time available. On occasion you will have the opportunity to receive visitors or leave the training center for a brief period.

After basic training, you either go to a technical training school where you are taught a specific skill or report directly to your duty station for on-the-job training.

JOB ASSIGNMENTS

The type of job you have and where you are stationed can range from being clerk at a base near your home to being missile technician aboard a ship in the western Pacific. Not only are there differences among the services themselves, but within any one military branch there is considerable variation depending on the needs of the service, your own qualifications, and your preferred location. On the average, you will spend from two to four years working in your military job. While you are learning a skill, you will advance in rate and assume more responsibility within the military structure. During this time, it is likely that you will be assigned to overseas duty and have the opportunity to visit other parts of the world.

The hundreds of different military occupations have been classified into twelve groups. It is estimated that about 75 percent of all enlisted jobs have a direct counterpart in the civilian workplace. Following are the twelve areas and examples of jobs within each.

Human Services
Caseworker, counselor, recreation specialist

Media and Public Affairs
Audiovisual specialist, photographer, reporter

Health Care
Dental assistant, nurse, laboratory technician

Engineering, Science, and Technical
Computer programmer, air traffic controller, weather observer

Administration
Recruiter, secretary, payroll specialist

Service
Cook, security policeman, fire fighter

Vehicle and Machinery Mechanics
Aircraft mechanic, heating and cooling mechanic, office machine repairer

Electronic and Electrical Equipment Repair
Electronic instrument repairer, power plant electrician, radio equipment repairer

Construction
Carpenter, ironworker, plumber

Machine Operator and Precision Work
Machinist, optician, truck driver

Transportation and Material Handling
Construction equipment operator, flight engineer, cargo handler

Combat Specialties
Infantryman, combat engineer, artillery crew member

ENLISTED RANKS

The following chart shows the rating system (equivalent to pay grades) for enlisted servicemembers and the common titles used by the different branches. Grades E-4 and above are designated NCOs—noncommissioned officers.

Pay Grade and Title

SERVICE	E-1	E-2	E-3	E-4	E-5	E-6	E-7	E-8	E-9
Army	Pvt	Pvt	Pvt 1st Class	Cpl	Sgt	Staff Sgt	Sgt 1st Class	Master Sgt	Sgt Major
Navy	Seaman Recruit	Seaman App	Seaman	PO 3rd Class	PO 2nd Class	PO 1st Class	Chief PO	Senior Chief PO	Master Chief PO
Marine Corps	Pvt	Pvt 1st Class	Lance Cpl	Cpl	Sgt	Staff Sgt	Gunnery Sgt	Master Sgt	Sgt Major
Air Force	Airman Basic	Airman	Airman 1st Class	Sgt	Staff Sgt	Tech Sgt	Master Sgt	Senior Master Sgt	Chief Master Sgt
Coast Guard	Seaman Recruit	Seaman App	Seaman	PO 3rd Class	PO 2nd Class	PO 1st Class	Chief PO	Senior Chief PO	Master Chief PO

Pvt = Private
Cpl = Corporal
Sgt = Sergeant
App = Apprentice
PO = Petty Officer

PROMOTIONS

While there is some variation from service to service, the typical advancement schedule looks like this:

Rate	Total Service Time
From E-1 to E-2	6 months
From E-2 to E-3	1 year
From E-3 to E-4	2 to 2 ½ years
From E-4 to E-5	3 ½ to 5 years
E-5 and above	Varies depending on performance and needs of the service

Promotion to E-2 and E-3 is fairly automatic, provided you have a satisfactory training record and receive a recommendation from your commanding officer.

Promotion to E-4 is a competitive process. It depends on your time in grade, and you must pass an advancement test and exhibit skill in your job specialty. Promotion to E-5 usually requires a high school diploma, further demonstration of job proficiency, and approval by a selection board.

PAY AND BENEFITS

These are the same for all five services. See Appendix E.

SPECIAL ENLISTMENT PROGRAMS

Each service offers a special track for highly qualified enlisted men and women. Those who have completed a Junior ROTC program in high school, have experience in critical job areas, or have completed a year or two of college have the greatest chance of being selected. These special programs offer such benefits as assignment to the job specialty of your choice, assignment to a specific geographic location, delay in reporting to active duty, an enlistment bonus of up to $8000, enlistment at E-2 or E-3 grade rather than as a recruit, and accelerated promotion.

JOINING THE RESERVES

While most enlisted servicemembers sign on for a three- to six-year term, an alternative is to join the National Guard or Reserve Forces and serve only about six months on active duty for training before joining a Reserve unit. As a reservist, you are required to drill one weekend a month and go on two weeks of active duty each year until you fulfill your required service obligation, usually five and a half years of drill participation and two years in the inactive, or nondrilling, Reserve.

You may hold down a regular civilian job while you participate in the National Guard or Reserves. Your military salary is based on the same schedule as an active-duty member's, but you are paid only for the days you attend drill or go on active duty.

As a reservist, you may qualify for education benefits. This is explained in Chapter 8.

WOMEN IN THE MILITARY

As mentioned previously, the role of women in the military has increased substantially during the past ten years, and the services actively encourage women to enlist. The number of women in the military now totals over 200,000, the highest number ever. With the exception of combat positions, nearly all of the jobs mentioned earlier are open to women as well as men.

Chapter 7

Earning College Credits on Active Duty

Although you may use the New G.I. Bill to pay for college courses you take while you are on active duty, it is much more common to save your benefits until you leave the service. Since the military has a number of its own in-service education programs that cost little or nothing, most servicemembers save the money they accumulate under the New G.I. Bill until they complete their tour of duty and enroll in college.

This chapter describes the programs offered by the services that enable you to earn college credit while you are on active duty. These fall into two categories: (1) courses you take on your own during off-duty hours and (2) credit you receive by demonstrating proficiency in your job specialty.

OFF-DUTY EDUCATION

Opportunities for off-duty education are organized in much the same way by each of the five military services. Each base has an education office that can assist you if you want to enroll in a local college or advance your education without attending class by taking correspondence courses.

For enlisted men and women who want to pursue a degree along traditional lines, the military has made arrangements with a number of colleges to offer flexible academic programs that take into account the time constraints and frequent reassignments of a servicemember. To accomplish this, the Army, Navy, Marine Corps, and Coast Guard have established an education network called Servicemembers Opportunity Colleges. The Air Force's program is called the Community College of the Air Force.

Arrangements for servicemembers to receive college credit through correspondence courses and examinations are coordinated by all the services under a program called Defense Activity for Non-Traditional Education Support.

In addition to making it easier for you to take courses, either in the classroom or through the mail, the services also provide money for your education in the form of tuition discounts. Under a program called Tuition Assistance, the military pays 75 percent of the tuition cost. (Depending on rank, time in service, and type of course, reimbursement can be as much as 90 percent.) You may take full advantage of Tuition Assistance while you are in the service and still receive benefits under the New G.I. Bill after you get out; there is no connection between the two programs.

Each service organizes its off-duty education programs in slightly different ways. The variations are described below.

Army

Within an overall framework called Army Continuing Education, the Army has divided its off-duty education into two levels called Servicemembers Opportunity Colleges Associate Degree (SOCAD) and Servicemembers Opportunity Colleges (SOC). Both are voluntary programs in which a servicemember takes courses—at a 75 percent discount—at nearby civilian colleges that participate in the SOCAD or SOC network.

You enroll in SOCAD if you want to work toward a two-year (associate) degree. You may take course work through any one of about fifty accredited colleges offering subjects in sixteen academic areas, ranging from accounting to computers to general studies. The colleges that participate in SOCAD have agreed to limit residency requirements, to grant credit for military work experience, to guarantee that course credit can be transferred to another college within the network, and to offer flexible class hours that take into account a servicemember's schedule. These special arrangements are designed to make it possible for you to earn an associate degree while you are in the service. Your degree is granted by the first college in which you enroll. If you are transferred to another base in the middle of your education program, you may enroll at another SOCAD college. The courses you take there will count toward your degree at your "home" institution.

SOC is set up in the same way as SOCAD, but it offers courses leading to a bachelor's degree or graduate degree. So far, the Army has been concentrating its resources on the SOCAD program, but it expects to expand its educational SOC services in the near future.

Navy

The Navy's commitment to the education of its enlisted men and women is coordinated within a structure called Navy Campus. Navy Campus hires civilian guidance counselors to help you design an education program that meets your goals.

The Navy's version of SOC is abbreviated SOCNAV. You can earn an associate or bachelor's degree at any one of about twenty-five SOCNAV colleges. These colleges have agreed to the same kind of flexible academic arrangements that were described for the Army's SOCAD program. SOCNAV participants are eligible for Tuition Assistance, which means you only have to pay 25 percent of the tuition cost. SOCNAV may eventually be extended to include graduate degrees.

The Navy also has a Certificate/Degree Program (offered primarily through the New York State Regents College) that works in much to same way as SOC-NAV, except that there is no residency requirement at all. The Regents College certifies work that you do, primarily by giving credit for your military job skills, thus making it possible for you to get a degree without attending class. This is particularly helpful for Navy personnel who serve on sea duty for extended periods of time.

The Navy Campus guidance counselor also can arrange for you to receive credit for courses taken at a college that doesn't participate in SOCNAV or the Certificate/Degree Program. Such opportunities are worked out individually and depend on your interests.

Another college option offered by the Navy is the Enlisted Education Advancement Program (EEAP). To be eligible, you must have a high school diploma and between four and fourteen years of active-duty service. If selected, you will attend a junior college full-time for two years and earn an associate degree. While in college, you will receive full pay and benefits, and you can use the New G.I. Bill to pay for educational costs. To be selected for EEAP, you will be required to extend your enlistment.

Finally, the Navy offers an educational opportunity called PACE (Program for Afloat College Education), in which college courses are taught aboard ships assigned to the Navy fleet.

Air Force

The Community College of the Air Force is a program that enables you to earn a two-year associate degree in applied science. You work toward this degree, which requires 64 semester hours, by combining on-the-job technical training or attendance at Air Force schools with college courses taken during off-duty hours. Counselors in education centers at the Air Force bases will help you enroll at a local college and take placement examinations, and they also will monitor your educational progress. Tuition Assistance covers 75 percent of the cost of Community College of the Air Force.

Education programs that lead to a bachelor's degree or graduate degree can also qualify for Tuition Assistance. All Air Force bases offer at least four separate subject areas at the bachelor's level and two graduate disciplines.

Marine Corps

The Marine Corps also uses a SOC arrangement to help you earn a college degree. SOC in the Marine Corps is structured along the same lines as in the Navy—you can take college-level courses and advance toward a degree under a flexible arrangement that allows for the time constraints of a Marine on active duty.

The Marine Corps also offers a program that lets you attend college full-time—the Degree Completion Program for Staff Non-Commissioned Officers. This arrangement makes it possible for sergeants who have previously started their college education to take up to eighteen months of leave to finish their degree. Although tuition is not covered, you receive full pay and allowances for your leave period.

Coast Guard

The Coast Guard also sponsors several off-duty education programs. As with the other branches of the Armed Forces, SOC is the framework under which most of the courses are offered. Depending on funding levels, it is possible to have 100 percent of your college tuition covered, an even more generous plan than in the other services.

In addition, the Coast Guard provides education in certain job specialties. For example, an Advanced Electronics Technology School enables senior enlisted servicemembers to pursue a college degree in electronics engineering, and the two-year Physician's Assistant Program leads to a medical certificate and a commission as a Chief Warrant Officer.

DANTES

The programs described above are designed to make it easy for you to enroll at a local college and take courses at reduced rates. It may be, however, that you can't arrange your work schedule so you can attend class. If this should be the case, you can further your education through correspondence courses and exams. Each of the five services participates in DANTES (Defense Activity for Non-Traditional Education Support). The DANTES program allows you to take tests to advance your level of education, demonstrate job proficiency, or qualify for college admission. The DANTES independent study catalog lists several hundred courses and examinations. Some of the major testing opportunities are:

- DANTES Subject Standard Tests (DSST). These are academic and vocational exams you take for college credit.

- College-Level Examination Program (CLEP). Through this program administered by the College Board, you can earn up to 30 semester hours of credit by passing five CLEP general exams.

- Proficiency Examination Program (PEP). Similar to CLEP, this is another credit-by-examination arrangement, sponsored by the American College Testing Program.

- Scholastic Aptitude Test (SAT) and American College Testing Program examination (ACT). These are used primarily to qualify for college admission, but they also are required for entry into officer training programs.

DANTES also includes tests that enable you to earn a high school diploma or have your job skills certified by professional associations.

When there is a fee for taking DANTES courses or examinations, it is usually possible to use the Tuition Assistance program to cover payments.

ON-THE-JOB TRAINING

The American Council on Education has worked out a system of evaluating how much college credit members of all five military branches can receive for attending a technical training school, performing certain job skills, or passing the Military Occupational Specialty exam. For example, a Navy communications specialist who serves three years on active duty can earn 10 to 15 credit hours through Navy schools, rating examinations, and professional experience.

The American Council on Education publishes the *Guide to the Evaluation of Educational Experiences in the Armed Services,* commonly known as the "green book," a listing of more than 400 military specialties for which it recommends that colleges give credit. To find out more about this method of receiving free college credit, talk with the education specialist on your base.

Chapter 8

Banking College Money While Enlisted: The New G.I. Bill

It has long been the practice of the U.S. government to reward men and women who serve in the Armed Forces by helping them pay for college after they leave the service. In one form or another this type of educational benefit has existed since World War II. The latest version is called the New G.I. Bill and is available to those who entered one of the five services—Army, Navy, Air Force, Marine Corps, or Coast Guard—after July 1, 1985.

The New G.I. Bill replaced the Veteran's Educational Assistance Plan (VEAP), the Armed Forces college financial aid program that had been in force previously. From the point of view of the servicemember, the New G.I. Bill offers a better education plan than VEAP, because, while the size of the college fund that you can accumulate is about the same, the government contributes a greater share of the total.

If you enlisted before July 1, 1985, you may be eligible to receive money for college under either VEAP or the Vietnam era G.I. Bill. Check with a Veterans Administration office to determine if you have benefits coming to you from one of these programs.

BENEFITS FOR ACTIVE-DUTY SERVICEMEMBERS

The basic principle of the New G.I. Bill is that you and the government work together to build an education savings account for you to use after you leave the service. To receive these particular benefits, you must sign up for a minimum of two years of continuous active duty. Although you are required to make a contribution to your college account, the military's share is far greater. As a result, the payoff to you is quite high. You should think seriously about participating in the New G.I. Bill when you enlist.

You may use your New G.I. Bill benefits for many different kinds of educational programs. You may enroll in an accredited college and work toward a two-year associate degree, a four-year bachelor's degree, or a graduate degree. You may also choose vocational or technical courses to prepare for such careers as welder, tractor/trailer driver, locksmith, secretary, television repairman, or computer programmer. In general, there are relatively few restrictions on the types of courses you can take, since veterans are encouraged to pursue whatever type of education will help them become productive members of society.

During your first year on active duty, you will automatically have a total of $1200 ($100 per month for twelve months) deducted from your paycheck to be put into your "education account." If you serve three years or more, the military adds $9600, for a total of $10,800. If you spend two years on active duty, the military contributes $7800, for a total of $9000. At the time you join the service, you may choose not to participate by "disenrolling." Unless you specify otherwise, the $100 payroll deduction will be made automatically each month.

The Veterans Administration will have to approve any program of study you decide to pursue later. If you attend college full-time, the $10,800 or $9000 is paid to you at the rate of $300 or $250 per month for up to thirty-six months. If you attend college less than full-time, your monthly benefits will be reduced proportionately. Although it is most common to use the New G.I. Bill to help pay for college expenses after you leave the military, it is possible to receive the benefits while still serving on active duty, provided you have served at least two years.

Your eligibility to receive benefits ends ten years after you leave the military. You will find a summary of the New G.I. Bill benefit schedule at the end of this chapter.

The New G.I. Bill also permits a servicemember to take a leave of absence for up to two years to pursue an educational program. To be eligible, you must complete your initial tour of duty and agree to extend your enlistment two years for each year of leave. During your leave period you continue to receive your basic pay while using your New G.I. Bill benefits.

The New Army College Fund

The law that created the New G.I. Bill permits a service to add money to the basic G.I. Bill in order to attract servicemembers into areas where there are personnel shortages. While the Navy, Air Force, Marine Corps, and Coast Guard have been able to attract enough volunteers to meet their recruiting quotas, the Army has had some difficulty in filling its critical jobs (about 25 percent of the total) and therefore has been authorized to establish the New Army College Fund, a program of educational benefits in addition to the New G.I. Bill.

The New Army College Fund provides a bonus of up to $400 per month over the basic G.I. Bill benefits. To be eligible, you must: (1) begin your military service after July 1, 1985; (2) graduate from high school; (3) score 50 or above on the Armed Forces Qualification Test; (4) enlist in an approved job specialty; (5) enroll in the New G.I. Bill; and (6) sign on for an active-duty tour of at least two years (reservists are not eligible).

The value of the New Army College Fund is $14,400 if you serve four years or more, $12,000 if you serve three years, and $8000 if you go on active duty for two years.

There are nearly 100 job specialties that can qualify you for the New Army College Fund, mostly in combat assignments such as Infantry, Field Artillery, and Armor or in technical areas such as electronics, air traffic control, and missile systems. Check with an Army recruiter for a current list.

Supplemental Benefits

The New G.I. Bill allows the services to pay supplemental benefits to servicemembers who stay in for more than five years (or two years of active duty and four years of Reserve duty) beyond the length of time needed to qualify for the basic plan. (At the time this book was printed, none of the services had implemented this option.)

BENEFITS FOR MEMBERS OF THE RESERVES

To receive assistance under the New G.I. Bill, it is not necessary for you to join the active-duty military for a minimum of two years. There are also payments for members of the Selected Reserve.

You are eligible if you enlist for a six-year term after July 1, 1985, provided you have completed your active-duty training and have a total of 180 days in the Reserves. If you already have a bachelor's degree you do not qualify.

As a full-time student you will receive $140 per month for up to thirty-six months, a total of $5040. The payment is reduced proportionately if you attend college less than full-time. The Reserve component of the New G.I. Bill does not require you to make a contribution—100 percent of the benefits are paid by the government.

Reservists may take the same kinds of courses as those described for active-duty servicemembers in the previous section on the New G.I. Bill.

SUMMARY OF THE NEW G.I. BILL

	Service-member's Contribution	Government's Contribution	Total Benefit	Army Bonus[1]	Total Army Benefit
Active Duty 4 years of service or longer	$1,200	$9,600	$10,800	$14,400	$25,200
3 years of service	1,200	9,600	10,800	12,000	22,800
2 years of service	1,200	7,800	9,000	8,000	17,000
Selected Reservists			5,040		

[1]See preceding page for eligibility for the Army bonus.

Chapter 9

Going from Enlisted Servicemember to Officer

The previous chapters in Part II have told you about enlisted education programs that lower the cost of earning college credits, either while you are in the military or after you get out. There are many reasons for wishing to pursue a college degree while you are in the service, an important one being to prepare yourself for a good job in the civilian work force. It is possible, however, that after a few years as an enlisted servicemember, you may decide that you would like to stay in the military and become an officer. To make such a transition possible, the military has set up a number of programs that will enable you to get your college degree and be commissioned upon graduation. One way to do this is to win an ROTC scholarship or enter a service academy through a special route that is reserved for enlisted servicemembers. The second way is to try for one of the special programs that each service offers to highly qualified enlisted men and women who want to become officers.

ROTC SCHOLARSHIPS

Each of the services that offers ROTC—the Army, the Navy (including the Marine Corps), and the Air Force—gives special consideration to enlisted men and women in the selection process.

There are some different eligibility rules for ROTC scholarship candidates who have prior enlisted service. You may request early release (up to thirty days) from active duty in order to enter an ROTC program. The experience that you have already gained in the military will count as credit toward the freshman and sophomore military courses, thus reducing your ROTC work load for the first two years. The age limit, 21 for high school applicants, can be increased by as many as four years, depending on the length of your prior service. The financial benefits are significant. In addition to the tuition assistance and $100 per month you receive from your ROTC scholarship, you also qualify for the New G.I. Bill benefits. With the ROTC scholarship paying tuition and the G.I. Bill taking care of room and board costs, it is likely that you will be able to attend college at very little cost.

SERVICE ACADEMIES

As an enlisted man or woman, you may apply to one of the service academies if you meet the basic eligibility standards and are medically qualified. Chapter 3 contains a description of the general admissions process at West Point, Annapolis (including the Marine Corps), the Air Force Academy, and the Coast Guard Academy.

There is, however, one important difference between the application procedure for an enlisted member of the active-duty military or Reserve Forces and that for a high school student. The three academies that require nominations (West Point, Annapolis, and the Air Force Academy) have each set aside 160 openings for members of the Armed Forces, thus making that process somewhat easier than it is for those who have no military affiliation. Furthermore, if it is determined that you have the potential to become an officer but need to improve your academic skills, you may be assigned to the academy preparatory school, where you will receive an intensive one-year course in English, science, and mathematics. If you do well in preparatory school, you have a good chance of receiving an appointment to an academy.

Since the Coast Guard Academy has neither a nomination procedure nor a preparatory school, an enlisted servicemember interested in attending the Coast Guard Academy follows the same admissions process as any other applicant.

SPECIAL PROGRAMS OF THE SERVICES

Navy

The Enlisted Commissioning Program is for well-qualified enlisted men and women who already have earned at least 30 college credits. If selected, you will be designated an officer candidate and will attend a six-week Naval Science Institute. You will then become affiliated with an NROTC unit and complete your studies leading to a bachelor's degree. While attending college, you will receive your enlisted pay and allowances. Although you must pay for tuition, you may use your benefits under the New G.I. Bill to cover other educational expenses. Upon graduation you will be commissioned as an ensign and report to active duty.

Marine Corps

The Marine Corps also offers an opportunity for selected enlisted men and women to become officers. Through the Marine Corps Enlisted Commissioning Program, enlisted servicemembers who qualify may attend a civilian college and work toward a bachelor's degree. After graduation, you attend Officer Candidate School and are commissioned as a second lieutenant upon completion of your program. You receive your regular pay and allowances while you attend college, but you must pay tuition. However, as in the Navy, the New G.I. Bill may be used for tuition and other expenses.

Air Force

Through its Airmen Education and Commissioning Program, the Air Force selects enlisted servicemembers who have already completed at least 45 semester hours of college and provides the means for them to finish their bachelor's degree. In this program, tuition and fees are paid by the Air Force. Upon graduation, you will be commissioned as a second lieutenant and assigned to active duty in a job specialty related to your undergraduate studies.

PART III
Appendixes

These appendixes contain lists of colleges with various ROTC arrangements. Although the information is accurate as of September 1989, the placement of ROTC detachments and the approval of cross-enrollment contracts can vary from year to year. Therefore, as you begin to make specific college plans, you should double-check with the service you are interested in to confirm all details.

The list of colleges in Appendixes A through D is current as of September 1989. It is naturally subject to change as new colleges sign ROTC agreements and others terminate their contracts.

Appendix A: Colleges Hosting Any ROTC Program

This list, arranged in alphabetical order by state, shows every college that has an ROTC host unit, which means that an ROTC detachment is located on campus. This appendix provides a complete overview of ROTC host activity in the United States.

Appendix B: Army ROTC Programs

This appendix shows the location of Army ROTC host colleges and extension centers. Host colleges, where the unit is actually located, are listed in the left-hand column. Extension centers, branches of a host ROTC unit that have been placed on another campus, are shown in the right-hand column next to the host college. Students attending a college with an ROTC extension center thus participate in ROTC on their own campus.

Under current Army ROTC rules, you must be enrolled at a host college or extension center to receive a four-year scholarship. If you enroll at a college that is neither a host college nor an extension center but that has a cross-enrollment agreement, you may be eligible for a two- or three-year in-college scholarship. Check at an Army ROTC unit for a list of colleges that have cross-enrollment agreements.

Appendix C: Naval ROTC Programs

This appendix lists colleges that host Naval ROTC programs or have a cross-enrollment arrangement. Students attending cross-enrollment colleges "commute" to the host unit indicated for their ROTC program. You may attend any college on the list and receive a two-, three-, or four-year Naval ROTC scholarship.

Appendix D: Air Force ROTC Programs

This appendix lists colleges that host Air Force ROTC programs or have a cross-enrollment arrangement. As with Navy cross-enrollment, this means that students attending cross-enrollment colleges "commute" to the host college indicated for their ROTC program. You may attend any of the colleges shown and be eligible for a two-, three-, or four-year Air Force ROTC scholarship. (If you attend a two-year college, you must agree to transfer to a four-year institution if your scholarship requires you to do so.)

Appendix E: Military Pay and Benefits

By combining base pay and typical supplemental pay, this table shows approximate salary levels for officers and enlisted servicemembers. The base pay rate is the same for each person in a particular category. Supplemental pay and allowances vary among individuals and take into account housing costs, meal expenses, assignment to areas with a high cost of living, and participation in hazardous duty.

Appendix F: Height and Weight Chart for Officers

This appendix is self-explanatory, setting down the required relationship between height and weight for officer candidates. The ranges vary somewhat among the different services.

Appendix A

Colleges Hosting Any ROTC Program

	Army	Navy	Air Force
Alabama			
Alabama Agricultural and Mechanical University, Normal	x		
Alabama State University, Montgomery			x
Auburn University, Auburn University	x	x	x
Jacksonville State University, Jacksonville	x		
Marion Military Institute, Marion	x		
Samford University, Birmingham			x
Troy State University, Troy			x
Tuskegee University, Tuskegee	x		x
University of Alabama, Birmingham	x		
University of Alabama, Tuscaloosa	x		x
University of North Alabama, Florence	x		
University of South Alabama, Mobile	x		
Alaska			
University of Alaska, Fairbanks	x		
Arizona			
Arizona State University, Tempe	x		x
Embry-Riddle Aeronautical University, Prescott			x
Northern Arizona University, Flagstaff	x		x
University of Arizona, Tucson	x	x	x
Arkansas			
Arkansas State University, State University	x		

	Army	Navy	Air Force
Arkansas (continued)			
Arkansas Tech University, Russellville	x		
Henderson State University, Arkadelphia	x		
Ouachita Baptist University, Arkadelphia	x		
Southern Arkansas University, Magnolia	x		
University of Arkansas, Fayetteville	x		x
University of Arkansas, Little Rock	x		
University of Arkansas, Pine Bluff	x		
University of Central Arkansas, Conway	x		
California			
California Polytechnic State University, San Luis Obispo	x		
California State University, Fresno	x		x
California State University, Long Beach	x		x
California State University, Sacramento			x
The Claremont Colleges, Claremont	x		
Loyola Marymount University, Los Angeles			x
San Diego State University, San Diego	x	x	x
San Jose State University, San Jose	x		x
Santa Clara University, Santa Clara	x		
University of California, Berkeley	x	x	x
University of California, Davis	x		
University of California, Los Angeles	x	x	x
University of California, Santa Barbara	x		
University of San Diego, San Diego		x	
University of San Francisco, San Francisco	x		
University of Southern California, Los Angeles	x	x	x
Colorado			
Colorado School of Mines, Golden	x		
Colorado State University, Fort Collins	x		x
Metropolitan State College, Denver	x		
University of Colorado, Boulder	x	x	x
University of Colorado, Colorado Springs	x		
University of Southern Colorado, Pueblo	x		
Connecticut			
University of Connecticut, Storrs	x		x

	Army	Navy	Air Force
Delaware			
University of Delaware, Newark	x		x
District of Columbia			
Georgetown University, Washington	x		
George Washington University, Washington		x	
Howard University, Washington	x		x
Florida			
Embry-Riddle Aeronautical University, Daytona Beach	x		x
Florida Agricultural and Mechanical University, Tallahassee	x	x	
Florida Institute of Technology, Melbourne	x		
Florida Southern College, Lakeland	x		
Florida State University, Tallahassee	x		x
Jacksonville University, Jacksonville		x	
Stetson University, DeLand	x		
University of Central Florida, Orlando	x		x
University of Florida, Gainesville	x	x	x
University of Miami, Coral Gables	x		x
University of South Florida, Tampa	x		x
University of Tampa, Tampa	x		
Georgia			
Augusta College, Augusta	x		
Columbus College, Columbus	x		
Fort Valley State College, Fort Valley	x		
Georgia Institute of Technology, Atlanta	x	x	x
Georgia Military College, Milledgeville	x		
Georgia Southern College, Statesboro	x		
Georgia State University, Atlanta	x		
Mercer University, Macon	x		
Morehouse College, Atlanta		x	
North Georgia College, Dahlonega	x		
Savannah State College, Savannah		x	
University of Georgia, Athens	x		x
Valdosta State College, Valdosta			x
Guam			
University of Guam, Mangilao	x		

	Army	Navy	Air Force
Hawaii			
University of Hawaii at Manoa, Honolulu	x		x
Idaho			
Boise State University, Boise	x		
Idaho State University, Pocatello	x		
University of Idaho, Moscow	x	x	
Illinois			
Chicago State University, Chicago	x		
Eastern Illinois University, Charleston	x		
Illinois Institute of Technology, Chicago		x	x
Illinois State University, Normal	x		
Knox College, Galesburg	x		
Loyola University of Chicago, Chicago	x		
Northern Illinois University, De Kalb	x		
Northwestern University, Evanston		x	
Parks College of Saint Louis University, Cahokia			x
Southern Illinois University, Carbondale	x		x
Southern Illinois University, Edwardsville			x
University of Illinois, Chicago	x		
University of Illinois, Urbana	x	x	x
Western Illinois University, Macomb	x		
Wheaton College, Wheaton	x		
Indiana			
Ball State University, Muncie	x		
Indiana State University, Terre Haute			x
Indiana University, Bloomington	x		x
Indiana University–Purdue University, Indianapolis	x		
Purdue University, West Lafayette	x	x	x
Rose-Hulman Institute of Technology, Terre Haute	x		
University of Notre Dame, Notre Dame	x	x	x
Iowa			
Iowa State University of Science and Technology, Ames	x	x	x
University of Iowa, Iowa City	x		x

	Army	Navy	Air Force
University of Northern Iowa, Cedar Falls	x		
Kansas			
Kansas State University, Manhattan	x		x
Pittsburg State University, Pittsburg	x		
University of Kansas, Lawrence	x	x	x
Wichita State University, Wichita	x		
Kentucky			
Eastern Kentucky University, Richmond	x		
Morehead State University, Morehead	x		
Murray State University, Murray	x		
University of Kentucky, Lexington	x		x
University of Louisville, Louisville	x		x
Western Kentucky University, Bowling Green	x		
Louisiana			
Grambling State University, Grambling			x
Louisiana State University and A&M College, Baton Rouge	x		x
Louisiana Tech University, Ruston			x
McNeese State University, Lake Charles	x		
Nicholls State University, Thibodaux	x		
Northeast Louisiana University, Monroe	x		
Northwestern State University of Louisiana, Natchitoches	x		
Southern Louisiana University, Hammond	x		
Southern University and A&M College, Baton Rouge	x	x	
Tulane University, New Orleans	x	x	x
University of Southwestern Louisiana, Lafayette			x
Maine			
Maine Maritime Academy, Castine		x	
University of Maine, Orono	x		x
University of Southern Maine, Portland	x		
Maryland			
Johns Hopkins University, Baltimore	x		

	Army	Navy	Air Force
Maryland (continued)			
Loyola College, Baltimore	x		
Morgan State University, Baltimore	x		
University of Maryland, College Park			x
Western Maryland College, Westminster	x		
Massachusetts			
Boston University, Boston	x	x	x
College of the Holy Cross, Worcester		x	x
Massachusetts Institute of Technology, Cambridge	x	x	x
Northeastern University, Boston	x		
University of Lowell, Lowell			x
University of Massachusetts, Amherst	x		x
Worcester Polytechnic Institute, Worcester	x		
Michigan			
Central Michigan University, Mount Pleasant	x		
Eastern Michigan University, Ypsilanti	x		
Michigan State University, East Lansing	x		x
Michigan Technological University, Houghton	x		x
Northern Michigan University, Marquette	x		
University of Detroit, Detroit	x		
University of Michigan, Ann Arbor	x	x	x
Western Michigan University, Kalamazoo	x		
Minnesota			
Bemidji State University, Bemidji	x		
College of St. Thomas, St. Paul			x
Mankato State University, Mankato	x		
Saint John's University, Collegeville	x		
University of Minnesota, Duluth			x
University of Minnesota, Twin Cities Campus, Minneapolis	x	x	x
Winona State University, Winona	x		
Mississippi			
Alcorn State University, Lorman	x		
Delta State University, Cleveland	x		
Jackson State University, Jackson	x		

	Army	Navy	Air Force
Mississippi State University, Mississippi State	x		x
Mississippi Valley State University, Itta Bena			x
University of Mississippi, University	x	x	x
University of Southern Mississippi, Hattiesburg	x		x

Missouri

	Army	Navy	Air Force
Central Missouri State University, Warrensburg	x		
Kemper Military School and College, Boonville	x		
Lincoln University, Jefferson City	x		
Missouri Western State College, St. Joseph	x		
Northeast Missouri State University, Kirksville	x		
Northwest Missouri State University, Maryville	x		
Southeast Missouri State University, Cape Girardeau			x
Southwest Missouri State University, Springfield	x		
University of Missouri, Columbia	x	x	x
University of Missouri, Rolla	x		x
Washington University, St. Louis	x		
Wentworth Military Academy and Junior College, Lexington	x		
Westminster College, Fulton	x		

Montana

	Army	Navy	Air Force
Montana State University, Bozeman	x		x
University of Montana, Missoula	x		

Nebraska

	Army	Navy	Air Force
Creighton University, Omaha	x		
Kearney State College, Kearney	x		
University of Nebraska, Lincoln	x	x	x
University of Nebraska, Omaha			x

	Army	Navy	Air Force
Nevada			
University of Nevada, Las Vegas	x		
University of Nevada, Reno	x		
New Hampshire			
University of New Hampshire, Durham	x		x
New Jersey			
New Jersey Institute of Technology, Newark			x
Princeton University, Princeton	x		
Rider College, Lawrenceville	x		
Rutgers University, New Brunswick	x		x
Saint Peter's College, Jersey City	x		
Seton Hall University, South Orange	x		
New Mexico			
Eastern New Mexico University, Portales	x		
New Mexico Military Institute, Roswell	x		
New Mexico State University, Las Cruces	x		x
University of New Mexico, Albuquerque		x	x
New York			
Canisius College, Buffalo	x		
Clarkson University, Potsdam	x		x
Cornell University, Ithaca	x	x	x
Fordham University, Bronx	x		
Hofstra University, Hempstead	x		
Manhattan College, Riverdale			x
Niagara University, Niagara University	x		
Polytechnic University, Brooklyn	x		
Rensselaer Polytechnic Institute, Troy	x	x	x
Rochester Institute of Technology, Rochester	x		x
St. Bonaventure University, St. Bonaventure	x		
St. John's University, Jamaica	x		
St. Lawrence University, Canton	x		
Siena College, Loudonville	x		
SUNY College, Brockport	x		
SUNY College, Fredonia	x		

	Army	Navy	Air Force
SUNY Maritime College, Bronx		x	
Syracuse University, Syracuse	x		x
University of Rochester, Rochester		x	

North Carolina

	Army	Navy	Air Force
Appalachian State University, Boone	x		
Campbell University, Buies Creek	x		
Davidson College, Davidson	x		
Duke University, Durham	x	x	x
East Carolina University, Greenville			x
Fayetteville State University, Fayetteville			x
North Carolina A&T State University, Greensboro	x		x
North Carolina State University, Raleigh	x		x
Saint Augustine's College, Raleigh	x		
University of North Carolina, Chapel Hill		x	x
University of North Carolina, Charlotte			x
University of North Carolina, Wilmington	x		
Wake Forest University, Winston-Salem	x		
Western Carolina University, Cullowhee	x		

North Dakota

	Army	Navy	Air Force
North Dakota State University, Fargo	x		x
University of North Dakota, Grand Forks	x		

Ohio

	Army	Navy	Air Force
Bowling Green State University, Bowling Green	x		x
Central State University, Wilberforce	x		
John Carroll University, University Heights	x		
Kent State University, Kent	x		x
Miami University, Oxford		x	x
Ohio State University, Columbus	x	x	x
Ohio University, Athens	x		x
University of Akron, Akron	x		x
University of Cincinnati, Cincinnati	x		x
University of Dayton, Dayton	x		
University of Toledo, Toledo	x		
Wright State University, Dayton			x

	Army	Navy	Air Force
Ohio (continued)			
Xavier University, Cincinnati	x		
Youngstown State University, Youngstown	x		
Oklahoma			
Cameron University, Lawton	x		
Central State University, Edmond	x		
East Central University, Ada	x		
Oklahoma State University, Stillwater	x		x
University of Oklahoma, Norman	x	x	x
Oregon			
Oregon State University, Corvallis	x	x	x
Portland State University, Portland	x		
University of Oregon, Eugene	x		
University of Portland, Portland			x
Pennsylvania			
Bucknell University, Lewisburg	x		
Carnegie Mellon University, Pittsburgh	x	x	
Clarion University of Pennsylvania, Clarion	x		
Dickinson College, Carlisle	x		
Drexel University, Philadelphia	x		
Duquesne University, Pittsburgh	x		
Gannon University, Erie	x		
Gettysburg College, Gettysburg	x		
Grove City College, Grove City			x
Indiana University of Pennsylvania, Indiana	x		
Lafayette College, Easton	x		
La Salle University, Philadelphia	x		
Lehigh University, Bethlehem	x		x
Pennsylvania State University, University Park	x	x	x
Saint Joseph's University, Philadelphia			x
Shippensburg University of Pennsylvania, Shippensburg	x		
Temple University, Philadelphia	x		
University of Pennsylvania, Philadelphia	x	x	

	Army	Navy	Air Force
University of Pittsburgh, Pittsburgh	x		x
University of Scranton, Scranton	x		
Valley Forge Military Junior College, Wayne	x		
Villanova University, Villanova		x	
Washington and Jefferson College, Washington	x		
Widener University, Pennsylvania Campus, Chester	x		
Wilkes College, Wilkes-Barre			x
Puerto Rico			
University of Puerto Rico, Mayagüez	x		x
University of Puerto Rico, Río Piedras	x		x
Rhode Island			
Providence College, Providence	x		
University of Rhode Island, Kingston	x		
South Carolina			
The Citadel, Charleston (men only)	x	x	x
Clemson University, Clemson	x		x
Furman University, Greenville	x		
Presbyterian College, Clinton	x		
South Carolina State College, Orangeburg	x		
University of South Carolina, Columbia	x	x	x
Wofford College, Spartanburg	x		
South Dakota			
South Dakota School of Mines and Technology, Rapid City	x		
South Dakota State University, Brookings	x		x
University of South Dakota, Vermillion	x		
Tennessee			
Austin Peay State University, Clarksville	x		
Carson-Newman College, Jefferson City	x		
East Tennessee State University, Johnson City	x		
Memphis State University, Memphis	x	x	x

	Army	Navy	Air Force
Tennessee (continued)			
Middle Tennessee State University, Murfreesboro	x		
Tennessee State University, Nashville			x
Tennessee Technological University, Cookeville	x		
University of Tennessee, Chattanooga	x		
University of Tennessee, Knoxville	x		x
University of Tennessee, Martin	x		
Vanderbilt University, Nashville	x	x	
Texas			
Angelo State University, San Angelo			x
Baylor University, Waco			x
East Texas State University, Commerce			x
Hardin-Simmons University, Abilene	x		
Midwestern State University, Wichita Falls	x		
Pan American University, Edinburg	x		
Prairie View A&M University, Prairie View	x	x	
Rice University, Houston		x	
St. Mary's University of San Antonio, San Antonio	x		
Sam Houston State University, Huntsville	x		
Southwest Texas State University, San Marcos			x
Stephen F. Austin State University, Nacogdoches	x		
Texas A&I University, Kingsville	x		
Texas A&M University, College Station	x	x	x
Texas Christian University, Fort Worth	x		x
Texas Tech University, Lubbock	x	x	x
Trinity University, San Antonio	x		
University of Houston, Houston	x		
University of North Texas, Denton			x
University of Texas, Arlington	x		
University of Texas, Austin	x	x	x
University of Texas, El Paso	x		x
University of Texas, San Antonio	x		x
West Texas State University, Canyon	x		

	Army	Navy	Air Force
Utah			
Brigham Young University, Provo	x		x
University of Utah, Salt Lake City	x	x	x
Utah State University, Logan	x		x
Weber State College, Ogden	x		
Vermont			
Norwich University, Northfield	x	x	x
Saint Michael's College, Winooski			x
University of Vermont, Burlington	x		
Virginia			
College of William and Mary, Williamsburg	x		
Hampton University, Hampton	x	x	
James Madison University, Harrisonburg	x		
Norfolk State University, Norfolk	x	x	
Old Dominion University, Norfolk	x	x	
University of Richmond, Richmond	x		
University of Virginia, Charlottesville	x	x	x
Virginia Military Institute, Lexington (men only)	x	x	x
Virginia Polytechnic Institute and State University, Blacksburg	x	x	x
Virginia State University, Petersburg	x		
Washington and Lee University, Lexington	x		
Washington			
Central Washington University, Ellensburg	x		x
Eastern Washington University, Cheney	x		
Gonzaga University, Spokane	x		
Seattle University, Seattle	x		
University of Puget Sound, Tacoma			x
University of Washington, Seattle	x	x	x
Washington State University, Pullman	x		x
West Virginia			
Marshall University, Huntington	x		
West Virginia State College, Institute	x		
West Virginia University, Morgantown	x		x

	Army	Navy	Air Force
Wisconsin			
Marquette University, Milwaukee	x	x	
Ripon College, Ripon	x		
St. Norbert College, De Pere	x		
University of Wisconsin, La Crosse	x		
University of Wisconsin, Madison	x	x	x
University of Wisconsin, Milwaukee	x		
University of Wisconsin, Oshkosh	x		
University of Wisconsin, Platteville	x		
University of Wisconsin, Stevens Point	x		
University of Wisconsin, Whitewater	x		
Wyoming			
University of Wyoming, Laramie	x		x

Army ROTC Programs

HOST COLLEGES	EXTENSION CENTERS
Alabama	
Alabama Agricultural and Mechanical University, Normal	
Auburn University, Auburn University	Auburn University, Montgomery
Jacksonville State University, Jacksonville	
Marion Military Institute, Marion	
Tuskegee University, Tuskegee	
University of Alabama, Birmingham	
University of Alabama, Tuscaloosa	
University of North Alabama, Florence	
University of South Alabama, Mobile	University of West Florida, Pensacola (FL)
Alaska	
University of Alaska, Fairbanks	
Arizona	
Arizona State University, Tempe	

HOST COLLEGES	EXTENSION CENTERS

Arizona (continued)

Northern Arizona University, Flagstaff

University of Arizona, Tucson

Arkansas

Arkansas State University,
 State University

Arkansas Tech University, Russellville

Henderson State University,
 Arkadelphia

Ouachita Baptist University,
 Arkadelphia

Southern Arkansas University,
 Magnolia

University of Arkansas, Fayetteville	Northeastern State University, Tahlequah (OK)

University of Arkansas, Little Rock

University of Arkansas, Pine Bluff

University of Central Arkansas, Conway

California

California Polytechnic State University,
 San Luis Obispo

California State University, Fresno

California State University, Long Beach

Claremont McKenna College, Claremont	California State Polytechnic University, Pomona California State University, Fullerton

HOST COLLEGES	EXTENSION CENTERS
	California State University, San Bernardino
San Diego State University, San Diego	
San Jose State University, San Jose	
Santa Clara University, Santa Clara	
University of California, Berkeley	
University of California, Davis	California State University, Chico California State University, Sacramento
University of California, Los Angeles	
University of California, Santa Barbara	
University of San Francisco, San Francisco	
University of Southern California, Los Angeles	

Colorado

Colorado School of Mines, Golden	
Colorado State University, Fort Collins	
Metropolitan State College, Denver	Mesa State College, Grand Junction
University of Colorado, Boulder	
University of Colorado, Colorado Springs	
University of Southern Colorado, Pueblo	

HOST COLLEGES	EXTENSION CENTERS
Connecticut	
University of Connecticut, Storrs	Central Connecticut State University, New Britain University of Bridgeport, Bridgeport
Delaware	
University of Delaware, Newark	Salisbury State University, Salisbury (MD)
District of Columbia	
Georgetown University, Washington	George Mason University, Fairfax (VA)
Howard University, Washington	Bowie State University, Bowie (MD)
Florida	
Embry-Riddle Aeronautical University, Daytona Beach	
Florida Agricultural and Mechanical University, Tallahassee	
Florida Institute of Technology, Melbourne	
Florida Southern College, Lakeland	
Florida State University, Tallahassee	
Stetson University, DeLand	
University of Central Florida, Orlando	
University of Florida, Gainesville	University of North Florida, Jacksonville
University of Miami, Coral Gables	

HOST COLLEGES	EXTENSION CENTERS
University of South Florida, Tampa	Saint Leo College, Saint Leo University of South Florida, St. Petersburg
University of Tampa, Tampa	

Georgia

Augusta College, Augusta	
Columbus College, Columbus	
Fort Valley State College, Fort Valley	Albany State College, Albany
Georgia Institute of Technology, Atlanta	Berry College, Rome
Georgia Military College, Milledgeville	
Georgia Southern College, Statesboro	Armstrong State College, Savannah
Georgia State University, Atlanta	
Mercer University, Macon	Georgia Southwestern College, Americus
North Georgia College, Dahlonega	
University of Georgia, Athens	

Guam

University of Guam, Mangilao	

Hawaii

University of Hawaii at Manoa, Honolulu	

Idaho

Boise State University, Boise	

HOST COLLEGES	EXTENSION CENTERS

Idaho (continued)

Idaho State University, Pocatello	
University of Idaho, Moscow	

Illinois

Chicago State University, Chicago	
Eastern Illinois University, Charleston	
Illinois State University, Normal	
Knox College, Galesburg	Bradley University, Peoria
Loyola University of Chicago, Chicago	
Northern Illinois University, De Kalb	
Southern Illinois University, Carbondale	Southeast Missouri State University, Cape Girardeau (MO)
University of Illinois, Chicago	
University of Illinois, Urbana	
Western Illinois University, Macomb	
Wheaton College, Wheaton	

Indiana

Ball State University, Muncie	
Indiana University, Bloomington	Indiana University Southeast, New Albany
Indiana University–Purdue University, Indianapolis	
Purdue University, West Lafayette	

HOST COLLEGES	EXTENSION CENTERS
Rose-Hulman Institute of Technology, Terre Haute	
University of Notre Dame, Notre Dame	

Iowa

Iowa State University of Science and Technology, Ames	Drake University, Des Moines
University of Iowa, Iowa City	
University of Northern Iowa, Cedar Falls	

Kansas

Kansas State University, Manhattan	
Pittsburg State University, Pittsburg	
University of Kansas, Lawrence	Emporia State University, Emporia
Wichita State University, Wichita	Fort Hays State University, Hays

Kentucky

Eastern Kentucky University, Richmond	Cumberland College, Williamsburg
Morehead State University, Morehead	
Murray State University, Murray	
University of Kentucky, Lexington	Kentucky State University, Frankfort
University of Louisville, Louisville	
Western Kentucky University, Bowling Green	

HOST COLLEGES	EXTENSION CENTERS
Louisiana	
Louisiana State University and A&M College, Baton Rouge	
McNeese State University, Lake Charles	Lamar University, Beaumont (TX)
Nicholls State University, Thibodaux	
Northeast Louisiana University, Monroe	Grambling State University, Grambling
Northwestern State University of Louisiana, Natchitoches	Centenary College of Louisiana, Shreveport Louisiana State University, Shreveport
Southeastern Louisiana University, Hammond	
Southern University and A&M College, Baton Rouge	
Tulane University, New Orleans	Dillard University, New Orleans
Maine	
University of Maine, Orono	
University of Southern Maine, Portland	
Maryland	
Johns Hopkins University, Baltimore	
Loyola College, Baltimore	
Morgan State University, Baltimore	
Western Maryland College, Westminster	

HOST COLLEGES	EXTENSION CENTERS

Massachusetts

Boston University, Boston	Stonehill College, North Easton
Massachusetts Institute of Technology, Cambridge	
Northeastern University, Boston	Salem State College, Salem Suffolk University, Boston
University of Massachusetts, Amherst	Western New England College, Springfield
Worcester Polytechnic Institute, Worcester	Fitchburg State College, Fitchburg

Michigan

Central Michigan University, Mount Pleasant

Eastern Michigan University, Ypsilanti

Michigan State University, East Lansing

Michigan Technological University, Houghton

Northern Michigan University, Marquette

University of Detroit, Detroit

University of Michigan, Ann Arbor

Western Michigan University, Kalamazoo

Minnesota

Bemidji State University, Bemidji

Mankato State University, Mankato

HOST COLLEGES	EXTENSION CENTERS

Minnesota (continued)

Saint John's University, Collegeville	St. Cloud State University, St. Cloud
University of Minnesota, Twin Cities Campus, Minneapolis	
Winona State University, Winona	

Mississippi

Alcorn State University, Lorman	
Delta State University, Cleveland	
Jackson State University, Jackson	
Mississippi State University, Mississippi State	
University of Mississippi, University	
University of Southern Mississippi, Hattiesburg	

Missouri

Central Missouri State University, Warrensburg	
Kemper Military School and College, Boonville	
Lincoln University, Jefferson City	
Missouri Western State College, St. Joseph	
Northeast Missouri State University, Kirksville	
Northwest Missouri State University, Maryville	

HOST COLLEGES	EXTENSION CENTERS
Southwest Missouri State University, Springfield	Missouri Southern State College, Joplin
University of Missouri, Columbia	
University of Missouri, Rolla	
Washington University, St. Louis	University of Missouri, St. Louis
Wentworth Military Academy and Junior College, Lexington	
Westminster College, Fulton	

Montana

Montana State University, Bozeman	Eastern Montana College, Billings
University of Montana, Missoula	

Nebraska

Creighton University, Omaha	University of Nebraska, Omaha
Kearney State College, Kearney	
University of Nebraska, Lincoln	

Nevada

University of Nevada, Las Vegas	
University of Nevada, Reno	

New Hampshire

University of New Hampshire, Durham	

New Jersey

Princeton University, Princeton	

HOST COLLEGES	EXTENSION CENTERS
New Jersey (continued)	
Rider College, Lawrenceville	
Rutgers University, New Brunswick	
Saint Peter's College, Jersey City	
Seton Hall University, South Orange	

New Mexico

Eastern New Mexico University, Portales	New Mexico Highlands University, Las Vegas
New Mexico Military Institute, Roswell	
New Mexico State University, Las Cruces	New Mexico Institute of Mining and Technology, Socorro University of New Mexico, Albuquerque

New York

Canisius College, Buffalo	
Clarkson University, Potsdam	
Cornell University, Ithaca	SUNY College, Cortland
Fordham University, Bronx	John Jay College of Criminal Justice, CUNY, New York Marist College, Poughkeepsie
Hofstra University, Hempstead	
Niagara University, Niagara University	
Polytechnic University, Brooklyn	
Rensselaer Polytechnic Institute, Troy	SUNY, Albany
Rochester Institute of Technology, Rochester	

HOST COLLEGES	EXTENSION CENTERS
St. Bonaventure University, St. Bonaventure	
St. John's University, Jamaica	
St. Lawrence University, Canton	
Siena College, Loudonville	
SUNY College, Brockport	
SUNY College, Fredonia	
Syracuse University, Syracuse	SUNY College, Oswego

North Carolina

Appalachian State University, Boone	
Campbell University, Buies Creek	
Davidson College, Davidson	University of North Carolina, Charlotte
Duke University, Durham	
North Carolina A&T State University, Greensboro	Elon College, Elon College
North Carolina State University, Raleigh	East Carolina University, Greenville
Saint Augustine's College, Raleigh	
University of North Carolina, Wilmington	
Wake Forest University, Winston-Salem	
Western Carolina University, Cullowhee	

North Dakota

North Dakota State University, Fargo	

HOST COLLEGES	EXTENSION CENTERS

North Dakota (continued)

University of North Dakota, Grand Forks	

Ohio

Bowling Green State University, Bowling Green	
Central State University, Wilberforce	
John Carroll University, University Heights	
Kent State University, Kent	
Ohio State University, Columbus	Franklin University, Columbus
Ohio University, Athens	Rio Grande College/Community College, Rio Grande
University of Akron, Akron	
University of Cincinnati, Cincinnati	
University of Dayton, Dayton	Wright State University, Dayton
University of Toledo, Toledo	
Xavier University, Cincinnati	Northern Kentucky University, Highland Heights (KY)
Youngstown State University, Youngstown	

Oklahoma

Cameron University, Lawton	
Central State University, Edmond	
East Central University, Ada	
Oklahoma State University, Stillwater	University of Tulsa, Tulsa

HOST COLLEGES	EXTENSION CENTERS
University of Oklahoma, Norman	

Oregon

Oregon State University, Corvallis	
Portland State University, Portland	
University of Oregon, Eugene	Oregon Institute of Technology, Klamath Falls

Pennsylvania

Bucknell University, Lewisburg	Bloomsburg University of Pennsylvania, Bloomsburg Mansfield University of Pennsylvania, Mansfield
Carnegie Mellon University, Pittsburgh	
Clarion University of Pennsylvania, Clarion	
Dickinson College, Carlisle	
Drexel University, Philadelphia	
Duquesne University, Pittsburgh	
Gannon University, Erie	Edinboro University of Pennsylvania, Edinboro
Gettysburg College, Gettysburg	Millersville University of Pennsylvania, Millersville Mount Saint Mary's College, Emmitsburg (MD) York College of Pennsylvania, York
Indiana University of Pennsylvania, Indiana	Slippery Rock University of Pennsylvania, Slippery Rock

HOST COLLEGES	EXTENSION CENTERS
Pennsylvania (continued)	
Lafayette College, Easton	East Stroudsburg University of Pennsylvania, East Stroudsburg
La Salle University, Philadelphia	
Lehigh University, Bethlehem	
Pennsylvania State University, University Park	Lock Haven University of Pennsylvania, Lock Haven Pennsylvania State University, Altoona Pennsylvania State University Delaware County Campus, Media Pennsylvania State University, Hazelton Pennsylvania State University, McKeesport Pennsylvania State University, Mont Alto Pennsylvania State University Ogontz Campus, Abington Pennsylvania State University, Schuylkill Haven Pennsylvania State University, The Behrend College, Erie
Shippensburg University of Pennsylvania, Shippensburg	
Temple University, Philadelphia	
University of Pennsylvania, Philadelphia	
University of Pittsburgh, Pittsburgh	
University of Scranton, Scranton	
Valley Forge Military Junior College, Wayne	

HOST COLLEGES	EXTENSION CENTERS
Washington and Jefferson College, Washington	California University of Pennsylvania, California
Widener University, Pennsylvania Campus, Chester	

Puerto Rico

University of Puerto Rico, Mayagüez	
University of Puerto Rico, Río Piedras	University of Puerto Rico, Cayey University of Puerto Rico, Humacao

Rhode Island

Providence College, Providence	Bryant College, Smithfield Rhode Island College, Providence
University of Rhode Island, Kingston	

South Carolina

The Citadel, The Military College of South Carolina, Charleston (men only)	
Clemson University, Clemson	
Furman University, Greenville	
Presbyterian College, Clinton	
South Carolina State College, Orangeburg	
University of South Carolina, Columbia	Benedict College, Columbia Francis Marion College, Florence
Wofford College, Spartanburg	

HOST COLLEGES	EXTENSION CENTERS
South Dakota	
South Dakota School of Mines and Technology, Rapid City	
South Dakota State University, Brookings	Northern State College, Aberdeen
University of South Dakota, Vermillion	
Tennessee	
Austin Peay State University, Clarksville	
Carson-Newman College, Jefferson City	
East Tennessee State University, Johnson City	
Memphis State University, Memphis	
Middle Tennessee State University, Murfreesboro	
Tennessee Technological University, Cookeville	
University of Tennessee, Chattanooga	
University of Tennessee, Knoxville	
University of Tennessee, Martin	
Vanderbilt University, Nashville	
Texas	
Hardin-Simmons University, Abilene	
Midwestern State University, Wichita Falls	
Pan American University, Edinburg	

HOST COLLEGES	EXTENSION CENTERS
Prairie View A&M University, Prairie View	
St. Mary's University of San Antonio, San Antonio	
Sam Houston State University, Huntsville	
Stephen F. Austin State University, Nacogdoches	
Texas A&I University, Kingsville	
Texas A&M University, College Station	Tarleton State University, Stephenville
Texas Christian University, Fort Worth	
Texas Tech University, Lubbock	
Trinity University, San Antonio	
University of Houston, Houston	
University of Texas, Arlington	Texas Woman's University, Denton
University of Texas, Austin	Southwest Texas State University, San Marcos
University of Texas, El Paso	
University of Texas, San Antonio	
West Texas State University, Canyon	

Utah

Brigham Young University, Provo

University of Utah, Salt Lake City

Utah State University, Logan

HOST COLLEGES	EXTENSION CENTERS

Utah (continued)

Weber State College, Ogden	

Vermont

Norwich University, Northfield	Dartmouth College, Hanover (NH)
University of Vermont, Burlington	

Virginia

College of William and Mary, Williamsburg	Christopher Newport College, Newport News
Hampton University, Hampton	
James Madison University, Harrisonburg	
Norfolk State University, Norfolk	Elizabeth City State University, Elizabeth City (NC)
Old Dominion University, Norfolk	
University of Richmond, Richmond	Longwood College, Farmville Virginia Commonwealth University, Richmond
University of Virginia, Charlottesville	
Virginia Military Institute, Lexington (men only)	
Virginia Polytechnic Institute and State University, Blacksburg	Radford University, Radford
Virginia State University, Petersburg	
Washington and Lee University, Lexington	Lynchburg College, Lynchburg

HOST COLLEGES	EXTENSION CENTERS

Washington

Central Washington University, Ellensburg	
Eastern Washington University, Cheney	
Gonzaga University, Spokane	
Seattle University, Seattle	
University of Washington, Seattle	
Washington State University, Pullman	Eastern Oregon State College, La Grande (OR)

West Virginia

Marshall University, Huntington	
West Virginia State College, Institute	
West Virginia University, Morgantown	Frostburg State University, Frostburg (MD)

Wisconsin

Marquette University, Milwaukee	
Ripon College, Ripon	
St. Norbert College, De Pere	
University of Wisconsin, La Crosse	
University of Wisconsin, Madison	
University of Wisconsin, Milwaukee	
University of Wisconsin, Oshkosh	
University of Wisconsin, Platteville	University of Dubuque, Dubuque (IA)
University of Wisconsin, Stevens Point	

HOST COLLEGES	EXTENSION CENTERS

Wisconsin (continued)

University of Wisconsin, Whitewater

Wyoming

University of Wyoming, Laramie

Appendix C

Naval ROTC Programs

HOST COLLEGES	CROSS-ENROLLMENTS
Alabama	
Auburn University, Auburn University	
Arizona	
University of Arizona, Tucson	Pima Community College, Tucson
California	
San Diego State University, San Diego	Point Loma Nazarene College, San Diego
	University of California, San Diego
University of California, Berkeley	California Maritime Academy, Vallejo
	California State University, Hayward
	California State University, Sacramento
	San Francisco State University, San Francisco
	San Jose State University, San Jose
	Santa Clara University, Santa Clara
	Stanford University, Stanford

HOST COLLEGES	CROSS-ENROLLMENTS
California (continued)	
	University of California, Davis
	University of San Francisco, San Francisco
University of California, Los Angeles	Biola University, La Mirada
	California State Polytechnic University, Pomona
	California State University, Fullerton
	California State University, Long Beach
	California State University, Los Angeles
	California State University, Northridge
	Loyola Marymount University, Los Angeles
	Northrop University, Los Angeles
	Occidental College, Los Angeles
	Pepperdine University, Malibu
	University of California, Irvine
University of San Diego, San Diego	
University of Southern California, Los Angeles	California Institute of Technology, Pasadena
	California State Polytechnic University, Pomona
	Claremont McKenna College, Claremont

Colorado

University of Colorado, Boulder	

District of Columbia

George Washington University, Washington	American University, Washington
	Catholic University of America, Washington

HOST COLLEGES	CROSS-ENROLLMENTS
	Georgetown University, Washington Howard University, Washington University of Maryland, College Park (MD) University of the District of Columbia, Washington

Florida

HOST COLLEGES	CROSS-ENROLLMENTS
Florida Agricultural and Mechanical University, Tallahassee	Florida State University, Tallahassee Tallahassee Community College, Tallahassee
Jacksonville University, Jacksonville	Florida Community College, Jacksonville University of North Florida, Jacksonville
University of Florida, Gainesville	

Georgia

HOST COLLEGES	CROSS-ENROLLMENTS
Georgia Institute of Technology, Atlanta	Agnes Scott College, Decatur Georgia State University, Atlanta Kennesaw State College, Marietta Oglethorpe University, Atlanta Southern College of Technology, Marietta
Morehouse College, Atlanta	Clark College, Atlanta Morris Brown College, Atlanta Spelman College, Atlanta
Savannah State College, Savannah	Armstrong State College, Savannah

Idaho

HOST COLLEGES	CROSS-ENROLLMENTS
University of Idaho, Moscow	Washington State University, Pullman (WA)

HOST COLLEGES	CROSS-ENROLLMENTS
Illinois	
Illinois Institute of Technology, Chicago	Lewis University, Romeoville University of Chicago, Chicago University of Illinois, Chicago
Northwestern University, Evanston	Loyola University of Chicago, Chicago North Park College, Chicago
University of Illinois, Urbana	Parkland College, Champaign
Indiana	
Purdue University, West Lafayette	
University of Notre Dame, Notre Dame	Bethel College, Mishawaka Holy Cross Junior College, Notre Dame Indiana University, South Bend Saint Mary's College, Notre Dame
Iowa	
Iowa State University of Science and Technology, Ames	
Kansas	
University of Kansas, Lawrence	
Louisiana	
Southern University and A&M College, Baton Rouge	Louisiana State University and A&M College, Baton Rouge
Tulane University, New Orleans	Dillard University, New Orleans Loyola University, New Orleans University of New Orleans, New Orleans Xavier University of Louisiana, New Orleans

HOST COLLEGES	CROSS-ENROLLMENTS
Maine	
Maine Maritime Academy, Castine*	University of Maine, Orono
Massachusetts	
Boston University, Boston	Boston College, Chestnut Hill Northeastern University, Boston
College of the Holy Cross, Worcester	Anna Maria College for Men and Women, Paxton Assumption College, Worcester Clark University, Worcester Worcester Polytechnic Institute, Worcester Worcester State College, Worcester
Massachusetts Institute of Technology, Cambridge	Harvard University, Cambridge Tufts University, Medford Wellesley College, Wellesley
Michigan	
University of Michigan, Ann Arbor	Eastern Michigan University, Ypsilanti
Minnesota	
University of Minnesota, Twin Cities Campus, Minneapolis	Augsburg College, Minneapolis Bethel College, St. Paul College of St. Thomas, St. Paul Concordia College, St. Paul Macalester College, St. Paul
Mississippi	
University of Mississippi, University	

*Marine Corps option not available.

HOST COLLEGES	CROSS-ENROLLMENTS

Missouri

University of Missouri, Columbia	Columbia College, Columbia

Nebraska

University of Nebraska, Lincoln	Concordia Teachers College, Seward Nebraska Wesleyan University, Lincoln

New Mexico

University of New Mexico, Albuquerque	

New York

Cornell University, Ithaca	Ithaca College, Ithaca SUNY College, Cortland
Rensselaer Polytechnic Institute, Troy	Bennington College, Bennington (VT) College of Saint Rose, Albany Hudson Valley Community College, Troy Russell Sage College, Troy Siena College, Loudonville SUNY, Albany Union College, Schenectady
SUNY Maritime College, Bronx*	Fordham University, Bronx
University of Rochester, Rochester	Monroe Community College, Rochester Nazareth College, Rochester Rochester Institute of Technology, Rochester St. John Fisher College, Rochester SUNY College, Brockport

*Marine Corps option not available.

HOST COLLEGES	CROSS-ENROLLMENTS
	SUNY College, Geneseo

North Carolina

Duke University, Durham	North Carolina Central University, Durham
University of North Carolina, Chapel Hill	North Carolina State University, Raleigh

Ohio

Miami University, Oxford	
Ohio State University, Columbus	Capital University, Columbus Otterbein College, Westerville

Oklahoma

University of Oklahoma, Norman	

Oregon

Oregon State University, Corvallis	Linn-Benton Community College, Albany

Pennsylvania

Carnegie Mellon University, Pittsburgh	
Pennsylvania State University, University Park	
University of Pennsylvania, Philadelphia	Bryn Mawr College, Bryn Mawr Drexel University, Philadelphia La Salle University, Philadelphia Saint Joseph's University, Philadelphia Swarthmore College, Swarthmore Temple University, Philadelphia

HOST COLLEGES	CROSS-ENROLLMENTS
Pennsylvania (continued)	
Villanova University, Villanova	

South Carolina

The Citadel, Charleston (men only)	
University of South Carolina, Columbia	

Tennessee

HOST COLLEGES	CROSS-ENROLLMENTS
Memphis State University, Memphis	Christian Brothers College, Memphis
Vanderbilt University, Nashville	Belmont College, Nashville David Lipscomb University, Nashville Fisk University, Nashville Tennessee State University, Nashville Trevecca Nazarene College, Nashville

Texas

HOST COLLEGES	CROSS-ENROLLMENTS
Prairie View A&M University, Prairie View	
Rice University, Houston	Houston Baptist University, Houston Texas Southern University, Houston University of Houston, Houston University of St. Thomas, Houston
Texas A&M University, College Station	
Texas Tech University, Lubbock	
University of Texas, Austin	St. Edward's University, Austin

HOST COLLEGES	CROSS-ENROLLMENTS

Utah

University of Utah, Salt Lake City	Weber State College, Ogden Westminster College, Salt Lake City

Vermont

Norwich University, Northfield	

Virginia

Hampton University, Hampton	
Norfolk State University, Norfolk	
Old Dominion University, Norfolk	
University of Virginia, Charlottesville	Piedmont Virginia Community College, Charlottesville
Virginia Military Institute, Lexington (men only)	
Virginia Polytechnic Institute and State University, Blacksburg	

Washington

University of Washington, Seattle	Seattle Pacific University, Seattle Seattle University, Seattle

Wisconsin

Marquette University, Milwaukee	
University of Wisconsin, Madison	

Appendix D
Air Force ROTC Programs

HOST COLLEGES	CROSS-ENROLLMENTS
Alabama	
Alabama State University, Montgomery	Auburn University, Montgomery Huntingdon College, Montgomery Troy State University, Montgomery
Auburn University, Auburn University	
Samford University, Birmingham	Birmingham-Southern College, Birmingham Miles College, Birmingham University of Alabama, Birmingham University of Montevallo, Montevallo
Troy State University, Troy	
Tuskegee University, Tuskegee	
University of Alabama, Tuscaloosa	
Arizona	
Arizona State University, Tempe	DeVry Institute of Technology, Phoenix Grand Canyon College, Phoenix

HOST COLLEGES	CROSS-ENROLLMENTS
Arizona (continued)	
Embry-Riddle Aeronautical University, Prescott	
Northern Arizona University, Flagstaff	
University of Arizona, Tucson	Pima Community College, Tucson

Arkansas

University of Arkansas, Fayetteville	

California

California State University, Fresno	West Coast Christian College, Fresno
California State University, Long Beach	Biola University, La Mirada California State Polytechnic University, Pomona California State University, Dominguez Hills, Carson California State University, Fullerton California State University, Los Angeles National University, Long Beach University of California, Irvine Whittier College, Whittier
California State University, Sacramento	National University, Sacramento University of California, Davis University of the Pacific, Stockton
Loyola Marymount University, Los Angeles	California State Polytechnic University, Pomona California State University, Dominguez Hills, Carson California State University, Fullerton

HOST COLLEGES	CROSS-ENROLLMENTS
	California State University, Long Beach California State University, Los Angeles California State University, Northridge California State University, San Bernardino Chapman College, Orange Northrop University, Los Angeles Pepperdine University, Malibu University of California, Riverside University of Redlands, Redlands Westmont College, Santa Barbara
San Diego State University, San Diego	National University, San Diego Point Loma Nazarene College, San Diego University of California, San Diego, La Jolla University of San Diego, San Diego
San Jose State University, San Jose	San Jose City College, San Jose Santa Clara University, Santa Clara Stanford University, Stanford
University of California, Berkeley	California State University, Hayward Holy Names College, Oakland Mills College, Oakland Saint Mary's College of California, Moraga Sonoma State University, Rohnert Park
University of California, Los Angeles	California Lutheran University, Thousand Oaks California State Polytechnic University, Pomona California State University, Dominguez Hills, Carson

HOST COLLEGES	CROSS-ENROLLMENTS

California (continued)

| | California State University, Fullerton
California State University, Long Beach
California State University, Los Angeles
California State University, Northridge
California State University, San Bernardino
Northrop University, Los Angeles
University of California, Irvine
University of California, Riverside
University of California, Santa Barbara
University of La Verne, La Verne |
| University of Southern California, Los Angeles | Biola University, La Mirada
California Institute of Technology, Pasadena
California Lutheran University, Thousand Oaks
California State Polytechnic University, Pomona
California State University, Dominguez Hills, Carson
California State University, Fullerton
California State University, Los Angeles
California State University, Northridge
California State University, San Bernardino
Chapman College, Orange
Claremont McKenna College, Claremont
Harvey Mudd College, Claremont
Northrop University, Los Angeles
Occidental College, Los Angeles |

HOST COLLEGES	CROSS-ENROLLMENTS

Pepperdine University, Malibu
Pomona College, Claremont
University of California, Irvine
University of California, Riverside
Whittier College, Whittier

Colorado

Colorado State University,
   Fort Collins

University of Colorado, Boulder   Metropolitan State College,
   Denver
Regis College, Denver
University of Colorado, Denver
University of Colorado Health
   Sciences Center, Denver

Connecticut

University of Connecticut, Storrs   Central Connecticut State
   University, New Britain
Eastern Connecticut State
   University, Willimantic
Southern Connecticut State
   University, New Haven
Trinity College, Hartford
University of Hartford,
   West Hartford
Wesleyan University, Middletown
Western Connecticut State
   University, Danbury
Yale University, New Haven

Delaware

University of Delaware, Newark   Delaware State College, Dover
Lincoln University,
   Lincoln University (PA)
Washington College,
   Chestertown (MD)
Wilmington College, New Castle

HOST COLLEGES	CROSS-ENROLLMENTS
District of Columbia	
Howard University, Washington	American University, Washington
	Catholic University of America, Washington
	Georgetown University, Washington
	George Washington University, Washington
	Trinity College, Washington
	University of the District of Columbia, Washington
Florida	
Embry-Riddle Aeronautical University, Daytona Beach	Bethune-Cookman College, Daytona Beach
	University of Central Florida, Daytona Beach
Florida State University, Tallahassee	Florida Agricultural and Mechanical University, Tallahassee
University of Central Florida, Orlando	
University of Florida, Gainesville	
University of Miami, Coral Gables	Barry University, Miami Shores
	Florida International University, Miami
	Florida Memorial College, Miami
	St. Thomas University, Miami
University of South Florida, Tampa	Florida Southern College, Lakeland
	Saint Leo College, Saint Leo
	University of Tampa, Tampa
Georgia	
Georgia Institute of Technology, Atlanta	Agnes Scott College, Decatur
	Clark College, Atlanta

HOST COLLEGES	CROSS-ENROLLMENTS
	Georgia State University, Atlanta Morehouse College, Atlanta Morris Brown College, Atlanta Southern College of Technology, Marietta Spelman College, Atlanta
University of Georgia, Athens	Medical College of Georgia, Augusta
Valdosta State College, Valdosta	

Hawaii

University of Hawaii at Manoa, Honolulu	Brigham Young University–Hawaii Campus, Laie Chaminade University, Honolulu Hawaii Loa College, Kaneohe Hawaii Pacific College, Honolulu University of Hawaii–West Oahu College, Pearl City

Illinois

Illinois Institute of Technology, Chicago	Chicago State University, Chicago Elmhurst College, Elmhurst Governors State University, University Park Lewis University, Romeoville Loyola University, Chicago North Central College, Naperville Northeastern Illinois University, Chicago Northern Illinois University, De Kalb North Park College, Chicago Northwestern University, Evanston Rush University, Chicago Saint Xavier College, Chicago University of Chicago, Chicago

HOST COLLEGES	CROSS-ENROLLMENTS

Illinois (continued)

	University of Illinois Medical Center, Chicago University of Illinois, Chicago
Parks College of Saint Louis University, Cahokia	Harris-Stowe State College, St. Louis (MO) Saint Louis University, St. Louis (MO) University of Missouri, St. Louis (MO) Washington University, St. Louis (MO)
Southern Illinois University, Carbondale	
Southern Illinois University, Edwardsville	McKendree College, Lebanon
University of Illinois, Urbana	

Indiana

Indiana State University, Terre Haute	Rose-Hulman Institute of Technology, Terre Haute
Indiana University, Bloomington	Butler University, Indianapolis DePauw University, Greencastle Indiana University–Purdue University, Indianapolis Marian College, Indianapolis Saint Mary-of-the-Woods College, Saint Mary-of-the-Woods
Purdue University, West Lafayette	
University of Notre Dame, Notre Dame	Indiana University, South Bend Saint Mary's College, Notre Dame

HOST COLLEGES	CROSS-ENROLLMENTS

Iowa

Iowa State University of Science and Technology, Ames	Drake University, Des Moines
University of Iowa, Iowa City	

Kansas

Kansas State University, Manhattan	
University of Kansas, Lawrence	MidAmerica Nazarene College, Olathe Washburn University, Topeka

Kentucky

University of Kentucky, Lexington	Centre College, Danville Eastern Kentucky University, Richmond Georgetown College, Georgetown Kentucky State University, Frankfort Transylvania University, Lexington
University of Louisville, Louisville	Bellarmine College, Louisville Indiana University Southeast, New Albany (IN) Spalding University, Louisville

Louisiana

Grambling State University, Grambling	
Louisiana State University and A&M College, Baton Rouge	Southern University and A&M College, Baton Rouge
Louisiana Tech University, Ruston	

HOST COLLEGES	CROSS-ENROLLMENTS

Louisiana (continued)

| Tulane University, New Orleans | Dillard University, New Orleans
Louisiana State University School
of Nursing, New Orleans
Loyola University, New Orleans
Our Lady of Holy Cross College,
New Orleans
Southern University,
New Orleans
University of New Orleans,
New Orleans
Xavier University of Louisiana,
New Orleans |
| University of Southwestern
Louisiana, Lafayette | |

Maine

| University of Maine, Orono | Colby College, Waterville
Husson College, Bangor |

Maryland

| University of Maryland,
College Park | Bowie State University, Bowie
George Mason University,
Fairfax (VA)
Johns Hopkins University,
Baltimore
Loyola College, Baltimore
Prince George's Community
College, Largo
Shepherd College,
Shepherdstown (WV)
Towson State University, Towson
Western Maryland College,
Westminster |

Massachusetts

| Boston University, Boston | Brandeis University, Waltham |

HOST COLLEGES	CROSS-ENROLLMENTS
	Northeastern University, Boston
	Simmons College, Boston
	Wentworth Institute of Technology, Boston
College of the Holy Cross, Worcester	Anna Maria College for Men and Women, Paxton
	Assumption College, Worcester
	Central New England College, Worcester
	Clark University, Worcester
	Worcester Polytechnic Institute, Worcester
	Worcester State College, Worcester
Massachusetts Institute of Technology, Cambridge	Harvard University, Cambridge
	Tufts University, Medford
	Wellesley College, Wellesley
University of Lowell, Lowell	Bentley College, Waltham
	Daniel Webster College, Nashua (NH)
	Endicott College, Beverly
	Gordon College, Wenham
	New England College, Henniker (NH)
	New Hampshire College, Manchester (NH)
	Notre Dame College, Manchester (NH)
	Rivier College, Nashua (NH)
	Saint Anselm College, Manchester (NH)
	Salem State College, Salem
University of Massachusetts, Amherst	Amherst College, Amherst
	Mount Holyoke College, South Hadley
	Smith College, Northampton
	Western New England College, Springfield

HOST COLLEGES	CROSS-ENROLLMENTS

Michigan

Michigan State University, East Lansing	
Michigan Technological University, Houghton	
University of Michigan, Ann Arbor	Concordia College, Ann Arbor Eastern Michigan University, Ypsilanti Lawrence Technological University, Southfield University of Michigan, Dearborn Wayne State University, Detroit

Minnesota

College of St. Thomas, St. Paul	Augsburg College, Minneapolis Bethel College, St. Paul College of St. Catherine, St. Paul Concordia College, St. Paul Hamline University, St. Paul Macalester College, St. Paul Northwestern College, St. Paul
University of Minnesota, Duluth	College of St. Scholastica, Duluth University of Wisconsin, Superior (WI)
University of Minnesota, Twin Cities Campus, Minneapolis	

Mississippi

Mississippi State University, Mississippi State	Mississippi University for Women, Columbus
Mississippi Valley State University, Itta Bena	Delta State University, Cleveland
University of Mississippi, University	

HOST COLLEGES	CROSS-ENROLLMENTS
University of Southern Mississippi, Hattiesburg	University of South Alabama, Mobile (AL) William Carey College, Hattiesburg

Missouri

Southeast Missouri State University, Cape Girardeau	
University of Missouri, Columbia	Columbia College, Columbia Stephens College, Columbia William Woods College, Fulton
University of Missouri, Rolla	

Montana

Montana State University, Bozeman	

Nebraska

University of Nebraska, Lincoln	Concordia Teachers College, Seward Doane College, Crete Nebraska Wesleyan University, Lincoln
University of Nebraska, Omaha	Bellevue College, Bellevue College of Saint Mary, Omaha Creighton University, Omaha Embry-Riddle Aeronautical University, Offutt Air Force Base University of Nebraska Medical Center, Omaha

New Hampshire

University of New Hampshire, Durham	Colby-Sawyer College, New London Keene State College, Keene

HOST COLLEGES	CROSS-ENROLLMENTS

New Hampshire (continued)

	Nathaniel Hawthorne College, Antrim
	New England College, Henniker
	New Hampshire College, Manchester
	Notre Dame College, Manchester
	Plymouth State College, Plymouth
	Rivier College, Nashua
	Saint Anselm College, Manchester
	University of Southern Maine, Portland (ME)

New Jersey

New Jersey Institute of Technology, Newark	Fairleigh Dickinson University, Teaneck
	Jersey City State College, Jersey City
	Kean College of New Jersey, Union
	Montclair State College, Upper Montclair
	Rutgers University, Newark
	Saint Peter's College, Jersey City
	Seton Hall University, South Orange
	Upsala College, East Orange
	William Paterson College of New Jersey, Wayne
Rutgers University, New Brunswick	Monmouth College, West Long Branch
	Princeton University, Princeton
	Rider College, Lawrenceville
	Rutgers University, Camden
	Trenton State College, Trenton
	Wagner College, Staten Island (NY)

HOST COLLEGES	CROSS-ENROLLMENTS

New Mexico

New Mexico State University,
   Las Cruces

University of New Mexico,
   Albuquerque

New York

Clarkson University, Potsdam	St. Lawrence University, Canton SUNY College, Potsdam
Cornell University, Ithaca	Ithaca College, Ithaca SUNY College, Cortland
Manhattan College, Riverdale	Adelphi University, Garden City College of Mount Saint Vincent, Riverdale Columbia University, New York Dowling College, Oakdale Long Island University, Brooklyn Campus, Brooklyn Long Island University, C. W. Post Campus, Brookville Mercy College, Dobbs Ferry Molloy College, Rockville Centre New York Institute of Technology, Old Westbury Pace University, New York Polytechnic University, Brooklyn St. Francis College, Brooklyn Heights St. Joseph's College, Suffolk Campus, Patchogue St. Thomas Aquinas College, Sparkill SUNY College, Old Westbury SUNY Maritime College, Bronx SUNY, Stony Brook

HOST COLLEGES	CROSS-ENROLLMENTS

New York (continued)

HOST COLLEGES	CROSS-ENROLLMENTS
Rensselaer Polytechnic Institute, Troy	Albany College of Pharmacy of Union University, Albany College of Saint Rose, Albany Maria College, Albany Russell Sage College, Troy Siena College, Loudonville Skidmore College, Saratoga Springs SUNY, Albany SUNY Empire State College, Saratoga Springs Union College, Schenectady
Rochester Institute of Technology, Rochester	Alfred University, Alfred Hobart and William Smith Colleges, Geneva Keuka College, Keuka Park Nazareth College, Rochester St. John Fisher College, Rochester SUNY College, Brockport SUNY College, Geneseo SUNY Empire State College, Saratoga Springs University of Rochester, Rochester Wells College, Aurora
Syracuse University, Syracuse	Le Moyne College, Syracuse New School for Social Research, New York SUNY College of Environmental Science and Forestry, Syracuse Utica College of Syracuse University, Utica

North Carolina

HOST COLLEGES	CROSS-ENROLLMENTS
Duke University, Durham	North Carolina Central University, Durham

HOST COLLEGES	CROSS-ENROLLMENTS
East Carolina University, Greenville	
Fayetteville State University, Fayetteville	Pembroke State University, Pembroke
North Carolina A&T State University, Greensboro	Bennett College, Greensboro Greensboro College, Greensboro Guilford College, Greensboro High Point College, High Point University of North Carolina, Greensboro
North Carolina State University, Raleigh	Meredith College, Raleigh Saint Augustine's College, Raleigh Shaw University, Raleigh
University of North Carolina, Chapel Hill	
University of North Carolina, Charlotte	Barber-Scotia College, Concord Belmont Abbey College, Belmont Davidson College, Davidson Johnson C. Smith University, Charlotte Queens College, Charlotte Wingate College, Wingate Winthrop College, Rock Hill (SC)

North Dakota

North Dakota State University, Fargo	Concordia College, Moorhead (MN) Moorhead State University, Moorhead (MN)

Ohio

Bowling Green State University, Bowling Green	Ohio Northern University, Ada University of Findlay, Findlay University of Toledo, Toledo
Kent State University, Kent	Mount Union College, Alliance

HOST COLLEGES	CROSS-ENROLLMENTS

Ohio (continued)

Miami University, Oxford	

Ohio State University, Columbus	Capital University, Columbus
	DeVry Institute of Technology,
	    Columbus
	Franklin University, Columbus
	Ohio Dominican College,
	    Columbus
	Ohio Wesleyan University,
	    Delaware
	Otterbein College, Westerville

Ohio University, Athens	

University of Akron, Akron	Ashland College, Ashland
	Baldwin-Wallace College, Berea
	Case Western Reserve University,
	    Cleveland
	Cleveland State University,
	    Cleveland

University of Cincinnati,
    Cincinnati
	Cincinnati Technical College,
	    Cincinnati
	College of Mount St. Joseph,
	    Mount St. Joseph
	Northern Kentucky University,
	    Highland Heights (KY)
	Thomas More College,
	    Crestview Hills (KY)
	Xavier University, Cincinnati

Wright State University,
    Dayton
	Antioch College, Yellow Springs
	Cedarville College, Cedarville
	Central State University,
	    Wilberforce
	University of Dayton, Dayton
	Urbana University, Urbana
	Wilberforce University,
	    Wilberforce
	Wilmington College, Wilmington

HOST COLLEGES	CROSS-ENROLLMENTS
	Wittenberg University, Springfield

Oklahoma

Oklahoma State University, Stillwater	
University of Oklahoma, Norman	Oklahoma Christian College, Oklahoma City Oklahoma City University, Oklahoma City

Oregon

Oregon State University, Corvallis	University of Oregon, Eugene Western Oregon State College, Monmouth
University of Portland, Portland	Concordia College, Portland Lewis and Clark College, Portland Oregon Health Sciences University, Portland Portland State University, Portland Warner Pacific College, Portland Williamette University, Salem

Pennsylvania

Grove City College, Grove City	Slippery Rock University of Pennsylvania, Slippery Rock
Lehigh University, Bethlehem	Allentown College of St. Francis de Sales, Center Valley Cedar Crest College, Allentown East Stroudsburg University of Pennsylvania, East Stroudsburg Kutztown University of Pennsylvania, Kutztown Lafayette College, Easton

HOST COLLEGES	CROSS-ENROLLMENTS

Pennsylvania (continued)

	Moravian College, Bethlehem Muhlenberg College, Allentown Pennsylvania State University, Allentown Campus, Fogelsville
Pennsylvania State University, University Park	

Saint Joseph's University,
    Philadelphia
	Bryn Mawr College, Bryn Mawr
	Cheyney University of
	    Pennsylvania, Cheyney
	Drexel University, Philadelphia
	Eastern College, Saint Davids
	Haverford College, Haverford
	La Salle University, Philadelphia
	Rutgers University, Camden (NJ)
	Swarthmore College, Swarthmore
	Temple University, Philadelphia
	Thomas Jefferson University,
	    Philadelphia
	University of Pennsylvania,
	    Philadelphia
	Villanova University, Villanova
	West Chester University of
	    Pennsylvania, West Chester
	Widener University, Pennsylvania
	    Campus, Chester

University of Pittsburgh,
    Pittsburgh
	Carlow College, Pittsburgh
	Carnegie Mellon University,
	    Pittsburgh
	Chatham College, Pittsburgh
	Duquesne University, Pittsburgh
	La Roche College, Pittsburgh
	Point Park College, Pittsburgh
	Robert Morris College, Coraopolis
	Saint Vincent College, Latrobe

Wilkes College, Wilkes-Barre	Bloomsburg University of Pennsylvania, Bloomsburg College Misericordia, Dallas

HOST COLLEGES	CROSS-ENROLLMENTS
	King's College, Wilkes-Barre
	Marywood College, Scranton
	University of Scranton, Scranton

Puerto Rico

HOST COLLEGES	CROSS-ENROLLMENTS
University of Puerto Rico, Mayagüez	Catholic University of Puerto Rico, Mayagüez
	Inter American University of Puerto Rico, San Germán
	University of Puerto Rico, Aguadilla Regional College, Ramey
University of Puerto Rico, Río Piedras	Bayamón Central University, Bayamón
	Bayamón Regional College, Río Piedras
	Inter American University of Puerto Rico, Metro Campus, Hato Rey
	Universidad Metropolitana, Río Piedras
	Universidad Politécnica de Puerto Rico, Hato Rey
	University of Puerto Rico, Bayamón Technical University College, Bayamón
	University of Puerto Rico, Carolina Regional College, Carolina
	University of Puerto Rico, Cayey University College, Cayey
	University of Puerto Rico, Humacao University College, Humacao
	University of the Sacred Heart, Santurce

South Carolina

The Citadel, Charleston (men only)

HOST COLLEGES	CROSS-ENROLLMENTS
South Carolina (continued)	
Clemson University, Clemson	Central Wesleyan College, Central
University of South Carolina, Columbia	Benedict College, Columbia Columbia College, Columbia

South Dakota

South Dakota State University, Brookings	

Tennessee

Memphis State University, Memphis	Christian Brothers College, Memphis LeMoyne-Owen College, Memphis Rhodes College, Memphis State Technical Institute, Memphis University of Tennessee Center for the Health Sciences, Memphis
Tennessee State University, Nashville	Belmont College, Nashville David Lipscomb University, Nashville Fisk University, Nashville Meharry Medical College, Nashville Middle Tennessee State University, Murfreesboro Trevecca Nazarene College, Nashville Vanderbilt University, Nashville Western Kentucky University, Bowling Green (KY)
University of Tennessee, Knoxville	

HOST COLLEGES	CROSS-ENROLLMENTS

Texas

Angelo State University, San Angelo	
Baylor University, Waco	Paul Quinn College, Waco University of Mary Hardin-Baylor, Belton
East Texas State University, Commerce	
Southwest Texas State University, San Marcos	Texas Lutheran College, Seguin
Texas A&M University, College Station	
Texas Christian University, Fort Worth	Baylor School of Nursing, Dallas Texas Wesleyan University, Fort Worth University of Texas, Arlington
Texas Tech University, Lubbock	Lubbock Christian University, Lubbock
University of North Texas, Denton	Southern Methodist University, Dallas Texas Woman's University, Denton University of Dallas, Irving
University of Texas, Austin	Concordia Lutheran College, Austin St. Edward's University, Austin
University of Texas, El Paso	
University of Texas, San Antonio	Trinity University, San Antonio

Utah

Brigham Young University, Provo	
University of Utah, Salt Lake City	Weber State College, Ogden

HOST COLLEGES	CROSS-ENROLLMENTS

Utah (continued)

	Westminster College, Salt Lake City
Utah State University, Logan	

Vermont

Norwich University, Northfield	
Saint Michael's College, Winooski	Lyndon State College, Lyndonville Trinity College, Burlington University of Vermont, Burlington

Virginia

University of Virginia, Charlottesville	
Virginia Military Institute, Lexington (men only)	
Virginia Polytechnic Institute and State University, Blacksburg	

Washington

Central Washington University, Ellensburg	
University of Puget Sound, Tacoma	Embry-Riddle Aeronautical University, Fort Lewis Pacific Lutheran University, Tacoma Saint Martin's College, Lacey Southern Illinois University, McChord Air Force Base
University of Washington, Seattle	Seattle Pacific University, Seattle Seattle University, Seattle

HOST COLLEGES	CROSS-ENROLLMENTS
Washington State University, Pullman	University of Idaho, Moscow (ID)

West Virginia

West Virginia University, Morgantown	Fairmont State College, Fairmont

Wisconsin

University of Wisconsin, Madison	Marquette University, Milwaukee Milwaukee School of Engineering, Milwaukee

Wyoming

University of Wyoming, Laramie	

Appendix E

Military Pay and Benefits

OFFICERS' YEARLY PAY

Grade/Rank

ARMY/AIR FORCE/MARINE CORPS	0-1 2nd Lieut	0-2 1st Lieut	0-3 Capt	0-4 Major	0-5 Lieut Colonel	0-6 Colonel	0-7 Brig General	0-8 Major General	0-9 Lieut General	0-10 General
NAVY/ COAST GUARD	Ensign	Lieut JG	Lieut	Lieut CDR	CDR	Capt	Rear Adm Lower Half	Rear Adm	Vice Adm	Adm
Years of Service										
< 2	$21,850	$24,844	$28,402	$31,218	$36,394	$43,480	$56,196	$65,500	$71,224	$79,018
2	22,538	26,550	30,910	36,190	41,116	46,824	59,304	67,144	72,814	81,430
3	26,022	30,618	32,546	38,044	43,316	49,260	59,304	68,490	74,140	81,430
4	26,022	31,434	35,244	38,044	43,316	49,260	59,304	68,490	74,140	81,430
6	26,022	31,960	36,588	38,594	43,316	49,260	61,492	68,490	74,140	81,430
8	26,022	31,960	37,642	39,926	43,316	49,260	61,492	72,814	75,760	85,986
10	26,022	31,960	39,292	42,082	44,346	49,260	64,448	72,814	75,760	85,986
14	26,022	31,960	41,710	45,600	48,706	50,608	67,144	75,760	78,478	85,986
20	26,022	31,960	41,710	48,306	55,440	60,604	77,102	85,986	85,986	85,986

1. These rates are accurate as of January 1, 1989.
2. There are pay raises at 12, 16, 18, and over 20 years that have not been included.
3. The yearly pay shown includes base pay and a typical food allotment and housing allowance. The food allotment and housing allowance can vary depending on such factors as whether base housing and meals are provided.
4. There is additional pay for serving in high-cost-of-living areas and hazardous-duty assignments.

Notes:
Adm = Admiral
Brig = Brigadier
Capt = Captain
CDR = Commander
JG = Junior Grade
Lieut = Lieutenant

ENLISTED YEARLY PAY

Grade

Years of Service	E-1	E-2	E-3	E-4	E-5	E-6	E-7	E-8	E-9
< 2	$13,656	$14,670	$15,194	$16,060	$17,348	$19,368	$21,800	—	—
2	13,656	14,670	15,730	16,638	18,330	20,506	22,978	—	—
3	13,656	14,670	16,144	17,272	18,918	21,058	23,568	—	—
4	13,656	14,670	16,570	18,184	19,472	21,694	24,144	—	—
6	13,656	14,670	16,570	18,436	20,340	22,252	24,726	—	—
8	13,656	14,670	16,570	18,436	20,916	22,816	25,288	$28,560	—
10	13,656	14,670	16,570	18,436	21,498	23,410	25,872	29,160	$33,072
14	13,656	14,670	16,570	18,436	22,348	24,822	27,334	30,312	34,224
20	13,656	14,670	16,570	18,436	22,348	25,692	28,766	32,044	35,964

Notes:
1. These rates are accurate as of January 1, 1989.
2. There are pay raises at 12, 16, 18, and over 20 years that have not been included.
3. The yearly pay shown includes base pay and a typical food allotment and housing allowance. The food allotment and housing allowance can vary depending on such factors as whether base housing and meals are provided.
4. There is additional pay for serving in high-cost-of-living areas and hazardous-duty assignments.

Officer and Enlisted Benefits

In addition to their pay, military personnel receive substantial benefits. These include:

- *Retirement.* You may retire after twenty years and receive 50 percent of your base pay for the rest of your life. You make no contribution to this retirement plan.

- *Medical Care.* You receive full medical and dental coverage at no charge. Most health-care costs for your family are also covered.

- *Life Insurance.* If you die while on active duty, your survivors are eligible for life insurance and other payments.

- *Commissary and Exchange Privileges.* You can shop in these military stores for food and other merchandise. On the average, costs are about 20 percent below those in civilian stores.

- *Education.* There are reduced rates for courses taken while on active duty, and the New G.I. Bill provides college money after you leave the service. Part II of this book explains these programs.

- *Social and Recreational Facilities.* Military bases provide child-care centers, movie theaters, golf courses, and similar facilities at costs considerably lower than the civilian average.

- *Travel.* You may travel for free on military passenger aircraft when there is space available.

- *Vacation.* You have thirty days of leave, or vacation, each year.

Appendix F

Height and Weight Chart for Officers

Following is a representative height and weight chart for military officer programs; it shows the standards used by the Air Force. Since there is some variation among requirements for the different branches, you should request specific information from the branch you are interested in.

Height (in inches)	Weight for Men (in pounds)	Weight for Women (in pounds)
58		90–120
59		92–122
60	100–158	94–124
61	102–163	96–128
62	103–168	98–130
63	104–174	100–132
64	105–179	102–135
65	106–185	104–138
66	107–191	106–141
67	111–197	109–145
68	115–203	112–150
69	119–209	115–154
70	123–215	118–158
71	127–221	122–162
72	131–227	125–167
73	135–233	128–171
74	139–240	130–175
75	143–246	133–179
76	147–253	136–184
77	151–260	139–188
78	153–267	141–192
79	159–273	144–196
80	166–280	147–201

MORE OUTSTANDING TITLES FROM PETERSON'S

PETERSON'S GUIDE TO FOUR-YEAR COLLEGES 1991

For twenty years, students, parents, and counselors have relied on *Peterson's Guide to Four-Year Colleges* because it's the *only* college guide that provides **all** the information they need to choose the college that's best for them.

Included in *Peterson's Guide to Four-Year Colleges 1991* are:

- Peterson's College Video Library
- Information on campus computer facilities
- Articles and essays on today's college trends

Profiling 1,950 accredited institutions that grant baccalaureate degrees in the United States and Canada, the guide includes the highest-quality and most accurate data on everything from admissions requirements to campus life.

In-Depth Descriptions of the Colleges—not found in any other college guide—go beyond the data profiles to provide a full, personal look at each college.

". . . the most current of guides . . ."
—The Washington Post

Coming in July 1990
21st Edition
$17.95 paperback
$33.95 hardcover

WINNING MONEY FOR COLLEGE
The High School Student's Guide to Scholarship Contests
Alan Deutschman

Winning Money for College is the only complete guide to scholarship competitions that students can enter and win on their own. It is the only compilation of facts, figures, dates, and advice pertaining to America's most prestigious—and most financially rewarding—scholarship competitions.

Awards totaling over $40-million are described. These national contests offer cash prizes for use at any college, with awards for everything from artistic talent to vocational skills.

The contest descriptions include eligibility requirements, the number and value of scholarships, how to enter, rules, deadlines, strategies, and interviews with previous winners. Samples of winning entries are included throughout.

The author: Alan Deutschman, a Princeton graduate, wrote this book after he won thousands of dollars through his own writing, public speaking, and academic performance.

". . . the high school students' guide to entering scholarship contests."
—USA Today

2nd Edition
$8.95 paperback

PETERSON'S COLLEGE MONEY HANDBOOK 1990

Anyone needing information about college costs and how to meet them needs this book. *Peterson's College Money Handbook* is the only book that gives a full account of costs and financial aid at more than 1,700 accredited four-year colleges in the United States.

With this handbook on expenses and available assistance at each college, parents, students, and guidance counselors can solve the mystery of college financial aid and make informed decisions.

The Guide is divided into four sections:

- An overview of the financial aid process
- A glossary that explains financial aid terms
- Cost and aid profiles of each college
- Directories that show which colleges offer each type of non-need aid and special money-saving options, indexed for easy reference

$18.95 paperback
$33.95 hardcover

Look for these and other Peterson's titles in your local bookstore

WITHDRAWAL